Galaxies and Quasars

Galaxies and Quasars

William J. Kaufmann, III

Department of Physics
San Diego State University

W. H. Freeman and Company
San Francisco

Sponsoring Editor: Arthur C. Bartlett; *Project Editor:* Pearl C. Vapnek; *Copyeditor:* Karen Judd; *Designer:* Marjorie Spiegelman; *Production Coordinator:* Linda Jupiter; *Illustration Coordinator:* Batyah Janowski; *Artist:* Dale Johnson; *Compositor:* Graphic Typesetting Service; *Printer and Binder:* The Maple-Vail Book Manufacturing Group.

Library of Congress Cataloging in Publication Data

Kaufmann, William J
      Galaxies and quasars.

      Bibliography: p.
      Includes index.
      1. Galaxies. 2. Quasars. I. Title.
QB857.K38   523.1'12   79-10570
ISBN 0-7167-1133-8
ISBN 0-7167-1134-6 pbk.

Printed in the United States of America

9 8 7 6 5 4 3 2 1

Cover photograph: © *1975 by AURA, Inc.*

to JoAnn H. with love

Contents

Preface

For thousands of years, people have gazed into the heavens and asked many of the same questions that we ask. How big is the universe? Where did it come from? Does the universe have an edge? A center? When did it begin? And will it ever come to an end?

As human beings, we are not satisfied with endless observations alone. It is not enough to go out night after night and simply make long lists of what we see in the sky. Instead, we want to know *why* things are the way they are. We want a *cosmology,* a theory from which we can understand the properties, evolution, and nature of the universe as a whole.

Ever since the dawn of civilization, every society and every religion has had a cosmology at the core of its teachings. In ancient times, these cosmologies were based primarily on divine revelation. The heavens were populated with gods and heroes, demons and monsters. The creation and behavior of the cosmos was seen as a direct result of supernatural forces.

Modern civilization really began when people no longer needed to view the universe as the stage on which gods and demons acted out their supernatural roles. Basic aspects of modern science and technology originated when we realized that the behavior of the universe could be entirely understood from the viewpoint of physical interactions between matter and radiation. Of course, this orientation removes much of the romanticism from the heavens. For example, the ancient Chinese believed that solar eclipses were caused by a terrible monster who devours the sun. During an eclipse, the citizenry were entreated to make a great commotion to scare off the monster, lest the sun be permanently devoured. Of course, their efforts were always rewarded. The modern explanation in terms of the moon's orbit about the earth does not seem quite as intriguing or fascinating.

Although a measure of fantasy and romanticism may have been lost, surely tremendous powers have been gained. With each new discovery about the cosmos, we have gained significant insights into the true nature of physical reality. These revelations have often had a direct translation into technology, thereby affecting the economies and politics of modern societies. Some of these changes are subtle, but others are profound. For example, the realization that thermonuclear reactions are occurring at the sun's center leads directly to the knowledge of how to build a hydrogen bomb.

Modern cosmology was born in the 1920s. Prior to that time, we did not know that galaxies exist. We did not know that we live in an expanding universe. Our current understanding of the structure of the universe is only half a century old. And virtually everything that everyone believed prior to that time is wrong. Prior to the 1920s, we simply did not have enough data to construct a complete picture.

But we have only begun to scratch the surface. Our efforts over the past few decades constitute a faltering glimpse as we peer for the first time across billions of light years. Quasars were discovered only twenty years ago. And the discovery of the cosmic microwave background is even more recent. Observations of the heavens at ultraviolet, infrared, and X-ray wavelengths are less than a decade old.

Surely our cosmology will be dramatically affected by this ongoing exploration of the distant universe. This exploration, which often leads us into totally unexpected domains, is one of the noblest quests of the human intellect. In the seventeenth century, Isaac Newton's ideas about gravity and his explanation of the orbits of the planets about the sun set the stage for the Industrial Revolution. An understanding of the evolution of galaxies or the nature of quasars could easily have an equally profound effect on the future course of civilization.

June 1979   William J. Kaufmann, III

Galaxies and Quasars

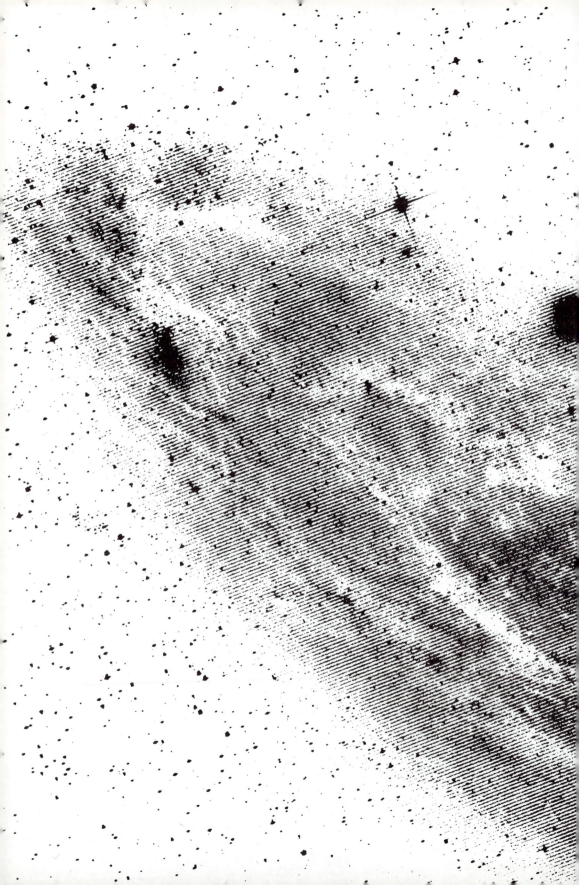

1

Exploring the Universe

It was said that they foretold the deaths of kings. Pestilence and plague were always predicted in the wake of these evil celestial omens. Misery and suffering would be long and hard. The devout crossed themselves and with bowed heads could be heard chanting "Dear God, protect us from the Turk, the Devil, and the Comet." The skies would surely be filled with comets on Judgment Day. Comets would herald the end of the world.

All of these medieval superstitions largely evaporated during the eighteenth century. In 1705, the English astronomer Edmund Halley published a book in which he demonstrated that comets are really members of our solar system. Halley was particularly intrigued by sightings of bright comets in 1531, 1607, and 1682. He argued that these sightings could be explained by a single comet that orbits the sun every 76 years. Halley did not live long enough to see the predicted return of his comet in 1758. But by then his ideas were accepted without question. In addition to the planets, our sun is orbited by numerous comets that occasionally produce some of the most spectacular and awe-inspiring sights in the sky.

Today we know that the solid portion of a comet consists of a frozen chunk of dust and ice that measures only a few kilometers across. A typical comet follows a highly elongated orbit that occasionally brings this frozen mixture into the inner regions of the solar system. As this interplanetary iceberg approaches the sun, the ices are vaporized. A large cloud of gases begins to develop. Blistering solar radiation causes these gases to glow, and the solar wind blows the shimmering material away from the comet's head into a long, billowing tail. *If* the comet's orbit happens to pass very near the sun (so that a large fraction of the ices are vaporized), and *if* the orbit also happens to pass near the earth (so that we can easily see the glowing gases), then we are treated to a dramatic sight that dominates the skies for many nights. An excellent example of a comet is shown in Figure 1-1.

Comet hunting became a popular fad during the late 1700s. Nearly everyone who had a decent telescope spent many long hours scanning the skies. With a little luck and a great deal of patience, it was possible to discover new members of our solar system. Prizes, medals, honors, and awards were heaped on the fortunate as-

Figure 1-1   A Comet
This beautiful comet dominated the skies during August 1957. Night after night it could be seen drifting slowly past the background stars. After a few weeks, the comet faded from view as it returned to the frigid depths of interplanetary space. (Hale Observatories.)

tronomer who happened to find a comet that produced a particularly spectacular sight.

But there were some problems. In order to be the discoverer of a comet, it is necessary to sight the object before it develops a long, conspicuous tail. The comet must be sighted while it is still quite faint. Astronomers therefore searched for faint, fuzzy objects in the sky that might be comets on their way toward the sun. But there are many faint, fuzzy objects across the sky that are *not* comets. These features are called *nebulas,* from the Latin word meaning "cloud." Nebulas proved to be very frustrating to the dedicated comet hunter.

To cope with this troublesome situation, the famous French comet hunters Charles Messier and Pierre Mechain made a list of 103 nebulas that were sometimes mistaken for distant comets. This list, called the *Messier Catalogue,* was published in 1781 (see the Appendix). The purpose of the list was to cut down the number of false alarms. Objects in the *Messier Catalogue* were to be ignored by comet hunters.

About half of the objects listed in the *Messier Catalogue* turned out to be clusters of stars. Some of these clusters are loose groupings of stars. These are called *open clusters.* Long-exposure photography through a telescope reveals that a typical open cluster contains several hundred stars, as shown in Figure 1-2.

Twenty-six open clusters were listed by Messier in his catalog. The remaining 29 clusters are called *globular clusters* because of their distinctive spherical appearance. A typical globular clusters, such as that shown in Figure 1-3, contains more than 100,000 stars.

Some of the objects on Messier's list actually are glowing clouds of interstellar gas and dust. These nebulas include the birthplaces of stars, such as the famous Orion Nebula (the 42nd object on Messier's list) shown in Figure 1-4. Other nebulas are stellar graveyards, such as the Crab Nebula (the 1st object on Messier's list) shown in Figure 1-5. There are about a dozen of these true nebulas on Messier's list. As with all of the objects in the *Messier Catalogue,* these nebulas look like tiny, faint, fuzzy patches of haze when viewed through a small telescope. The dramatic pictures on these pages are

6

Figure 1-2   The Open Cluster M67 (also called NGC 2682)
About two dozen open star clusters are listed in the Messier
Catalogue. *This open cluster (the 67th object on Messier's list) is
located in the constellation of Cancer and contains several
hundred loosely grouped stars. (Kitt Peak National Observatory.)*

7

Figure 1-3   The Globular Cluster M5 (also called NGC 5904)
More than two dozen globular star clusters are listed in the Messier
Catalogue. *This globular cluster (the 5th object on Messier's list) is
located in the constellation of Serpens and contains more than
100,000 densely grouped stars. (Kitt Peak National Observatory.)*

obtained by long-exposure photography through some of the world's largest telescopes.

All of these objects — star clusters and true nebulas — make up about two-thirds of Messier's list. They were never terribly controversial. Gradually, as data accumulated over the years, a giant picture began to emerge from which all of these objects could be understood and appreciated. For example, as we shall see in a later chapter, the locations of star clusters revealed the structure of our Milky Way Galaxy. And as astrophysicists began to understand the life cycles of stars, we learned that some nebulas are stellar embryos while others are stellar corpses.

But about a third of the objects in the *Messier Catalogue* are neither star clusters nor true nebulas. Although they were called "nebulas" for almost two centuries, these objects are definitely not glowing clouds of interstellar gas. These mysterious "nebulas," such as the Andromeda "Nebula" shown in Figure 1-6, became extremely controversial. For many years they were the subject of heated debates and vigorous arguments. The final resolution was destined to give humanity profound insight into the true nature of the universe as a whole.

About the same time that Messier was compiling his famous list, a German-born musician became interested in astronomy. William Herschel had moved from Hanover to London as a young man and at the age of 35 happened to purchase an astronomy book. He was so enthralled by what he read that he decided to build several large telescopes for himself. Indeed, Herschel probably spent more time looking through a telescope than anyone before him. His diligence was rewarded in 1781, when, at the age of 43, he accidentally discovered Uranus, the seventh planet from the sun.

William Herschel is responsible for some of the most important advances in astronomy during the late 1700s and early 1800s. For example, he was the first person to try to discover our location within the Milky Way Galaxy. Ever since Galileo first pointed his telescope toward the skies, it was realized that the hazy band of light called the Milky Way actually consists of millions of very faint stars. A portion of the Milky Way is shown in Figure 1-7. Because the Milky Way extends completely around the sky in one continuous encircling

Figure 1-4   The Orion Nebula (called M42 or NGC 1976)
*Many young stars are scattered throughout this huge cloud of gas
in the constellation of Orion. Intense ultraviolet radiation from
these young stars causes the gas to shine. This nebulosity (the 42nd
object on Messier's list) is just barely visible to the naked eye as a
faint, fuzzy "star" in Orion's sword. (Lick Observatory.)*

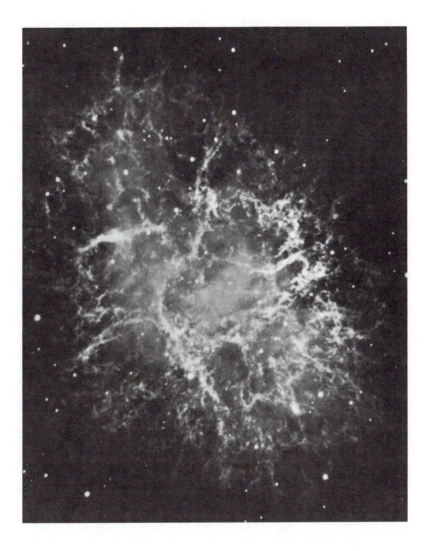

Figure 1-5 The Crab Nebula (called M1 or NGC 1952)
This beautiful nebulosity (the 1st object on Messier's list) is the corpse of a star. Nine hundred years ago, a star ended its life with a violent supernova explosion. The stellar remnant of this cataclysm is located in the constellation of Taurus. (Lick Observatory.)

***Figure 1-6   The Andromeda "Nebula" (called M31 or
NGC 224)***
*This photograph by an amateur astronomer is a good represen-
tation of what you see when you look through a moderate-
sized telescope at the 31st object on Messier's list. It is not a simple
cluster of stars like M5 or M67. Neither is it a real nebula like M1
or M42. For a century and a half, astronomers were baffled by views
like this. (Courtesy of Jo Ann Hunt.)*

12

band, it is reasonable to suppose that the sun is one star among millions in a huge, disk-shaped aggregation. This vast assemblage of stars is called the *Milky Way Galaxy,* or simply the Galaxy for short. Herschel reasoned that he should be able to deduce our location inside this enormous disk of stars by simply counting the numbers of stars in various directions. In 1785, he published the results of counting stars in 683 selected regions of the sky and concluded that we are at the center of the Galaxy. Although his method and results were fraught with errors, Herschel's ideas were certainly dramatic and innovative for the times. Herschel's map of the Galaxy is shown in Figure 1-8.

While scanning the skies, Herschel realized that hundreds upon hundreds of faint, fuzzy nebulas are scattered in almost every direction of space. Messier had listed only the brightest hundred that were sometimes mistaken for distant comets. But instead of ignoring these nebulas, Herschel became intrigued and resolved to do a more comprehensive search. Within seven years he had discovered and catalogued 2,000 previously unnoticed nebulas.

It took decades to cover the entire sky. William Herschel's son, John, vigorously continued where his father had left off. In 1864, John Herschel published *The General Catalogue of Nebulae,* which listed 5,079 objects. In 1888, seventeen years after John Herschel's death, this catalog was revised and enlarged by John Dreyer to include 7,840 nebulas and clusters. Dreyer's final version, *The New General Catalogue,* was published in 1895. This was followed shortly by two supplements, the *Index Catalogues.* Altogether an incredible total of nearly 15,000 nebulas and clusters were listed and described. And this was all done visually, without the aid of photography—a monumental achievement!

The New General Catalogue and the *Index Catalogues* are so comprehensive and complete that modern astronomers commonly use NGC and IC numbers when referring to nebulas and clusters in the sky. Thus, for example, the Crab Nebula is known as NGC 1952 because it is the 1,952nd entry on Dreyer's list.

As with Messier's list, the NGC and IC's contain many star clusters, many true nebulas, and a host of controversial nonstellar objects similar to the Andromeda "Nebula." Back as far as 1750, the

Figure 1-7   The Milky Way

This mosaic of wide-angle photographs shows the Milky Way from Cassiopeia (on the left) to Sagittarius (on the right). The Milky Way actually consists of a band of faint stars and stretches all the way around the sky. (Hale Observatories.)

Figure 1-8   Herschel's Map of the Galaxy

By laboriously counting the number of stars in various regions across the sky, William Herschel tried to discover the shape and size of the Galaxy. This map was the final result. Herschel believed that the sun was at the center of the Galaxy. (Yerkes Observatory.)

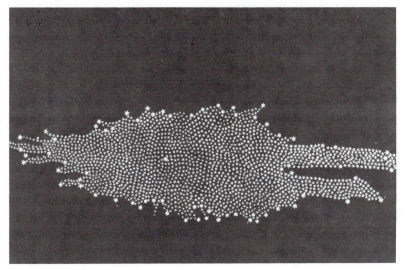

Englishman Thomas Wright speculated that some of these "nebulas" might be separate huge systems of stars like our own Milky Way Galaxy. Wright's ideas were largely ignored by almost everyone except the famous German philosopher Immanuel Kant. In 1755, Kant extended Wright's ideas by arguing in favor of "island universes," huge rotating disks of millions upon millions of stars widely scattered across space. Kant even proposed that the Andromeda "Nebula" is a good example, a vast stellar system virtually identical to the Milky Way Galaxy. Kant was also largely ignored.

William Parsons was the third Earl of Rosse. He was rich, he liked machines, and he was fascinated by astronomy. Accordingly, Lord Rosse set about the business of building gigantic telescopes. By February 1845, his *pièce de resistance* was finished. The telescope's massive primary mirror measured 6 feet in diameter. It was mounted at one end of a 60-foot tube that was controlled by cables, straps, pulleys, and cranes. For a brief period of time, Lord Rosse's contraption enjoyed the dubious reputation of being the largest (and most dangerous!) telescope in the world. During this reign, Lord Rosse discovered that many of Herschel's nebulas could be resolved into numerous closely packed stars. This proved that many of Herschel's "nebulas" are actually star clusters. Equally important, Lord Rosse found that M51 (the 51st object on Messier's list) exhibited a distinctly spiral structure. His sketch of M51 is shown in Figure 1-9. From the characteristic appearance of M51, Lord Rosse concluded that this nebula is actually a huge rotating whirlpool of stars. Indeed, this object is today often called the "Whirlpool Galaxy." A superb modern photograph of M51 is shown in Figure 1-10.

The invention of photography was one of the most important technological advances for astronomers during the late 1800s. No longer was it necessary to rely on crude, hand-drawn sketches. Astronomers could obtain accurate, permanent records of what they saw through their telescopes. Perhaps more important, astronomers soon discovered that enormous amounts of fine detail could be photographed simply by exposing a photographic plate for an extended period of time at the telescope's focus. This technique of astrophotography was pioneered by Isaac Roberts, an energetic amateur astronomer in Britain. In 1888, Roberts succeeded in taking a photograph of M31 that revealed spiral structure. Four views of M31 are shown in Figure 1-11. Objects such as M51 and M31 became known as "spiral nebulas."

In 1908, construction on the 60-inch telescope on Mount Wilson near Pasadena, California, was completed. Although this telescope was slightly smaller than Lord Rosse's colossus, it was a much finer piece of machinery. And astronomers began taking photographs of objects all across the sky. Within a decade, the Mount Wilson 100-inch telescope was finished, and the quality of astronomical photographs increased still further. Paradoxically, the dilemma surrounding the spiral nebulas became even more perplexing.

Some astronomers liked Kant's idea that these mysterious spiral nebulas were huge, distant galaxies. After all, there is nothing very special about our planet. Nine planets orbit our sun alone, and presumably there are many other planetary systems orbiting many other stars. There is also nothing very special about our star. The sun is just one of millions upon millions of stars that can be seen through telescopes. So perhaps there is nothing very special about our Galaxy. Perhaps the Milky Way Galaxy is just one of thousands of galaxies scattered across the universe.

But the majority opinion seemed to lean in the other direction. Perhaps these spiral nebulas are just little whirlpools of stars and gas at relatively nearby distances. Perhaps these spiral nebulas are scattered around the Milky Way Galaxy just like all the other star clusters and nebulas in Messier's list and in the NGC. Indeed, there seemed to be some observational evidence favoring this position. One astronomer erroneously believed that he could actually see the rotation of the spiral nebula by comparing photographs taken a few years apart. It was realized that this rotation could only be detected if the spiral nebulas were nearby. Other astronomers discovered exploding stars in some of the spiral nebulas. These exploding stars were mistaken to be *novas* (rather than *supernovas,* which are millions of times brighter). Distance estimates based on this mistaken identity placed the spiral nebula around the Milky Way.

The controversy came to a head in April 1920, when astronomers squared off in a debate sponsored by the National Academy of Sciences in Washington, D.C. Harlow Shapley of Mount

17

Figure 1-9   Lord Rosse's Sketch of M51
*Using a large telescope of his own design, Lord Rosse in England
was able to distinguish spiral structure in this "nebula."
(Reproduced courtesy of Lund Humphries.)*

Figure 1-10   The Spiral Galaxy M51 (also called NGC 5194)
This spiral galaxy in the constellation of Canes Venatici is often called the "Whirlpool Galaxy" because of its distinctive appearance. The "blob" at the end of one of the spiral arms is a small companion galaxy called NGC 5195. (Kitt Peak National Observatory.)

1-minute exposure   *5-minute exposure*

30-minute exposure   *45-minute exposure*

Figure 1-11   Photography of the Andromeda "Nebula"
Time exposure photography records details that are invisible to the human eye. These four views of M31 show how the amount of detail increases dramatically in proportion to exposure time. (Kitt Peak National Observatory.)

Wilson Observatory argued that the spiral nebulas are nearby. He was opposed by Heber D. Curtis of Lick Observatory, who championed the more distant interpretation. In retrospect, we now realize that Shapley used all the right arguments to come to the wrong conclusion, whereas Curtis used many wrong arguments to arrive at the right conclusion.

Little public interest was shown in this historic Shapley-Curtis debate. In spite of the cosmic questions at issue, the Washington press was largely indifferent. Perhaps that was for the best, because no firm conclusions could be reached. Astronomers were grappling with insufficient data and disputable observations. But while this debate raged, a young lawyer took up the study of astronomy. He would soon make a series of discoveries and observations with which no one could quarrel. His insights would settle the issue once and for all and point the way toward revealing the structure of the cosmos.

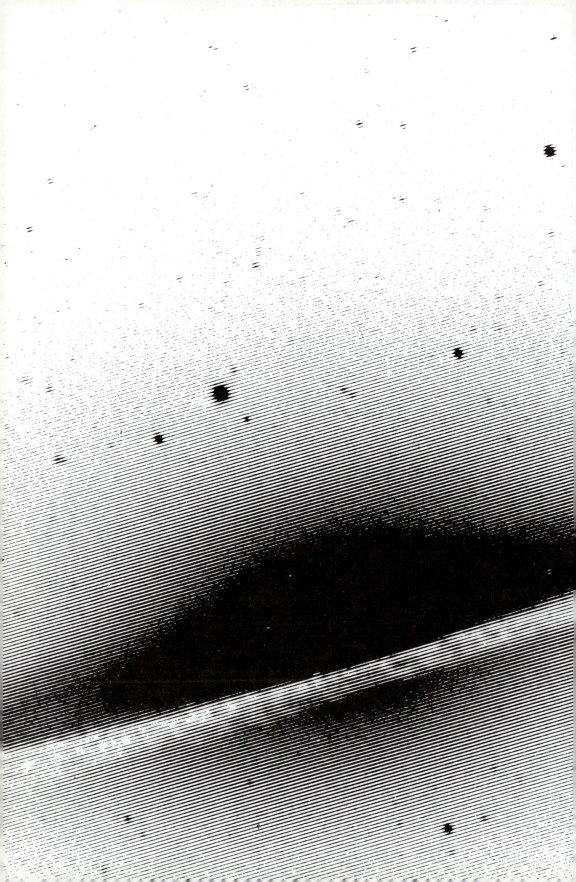

2

Discovering the Galaxies

There are many different kinds of stars in the sky. Most of the stars that you can see on a clear night are very similar to the sun. Like the sun, they are fairly young stars that are shining because a simple thermonuclear reaction called *hydrogen burning* is occurring at their centers.

But there are other kinds of stars. For example, almost every reddish-colored star you see in the night sky is a red giant. Antares in Scorpius, Arcturus in Boötes, and Betelgeuse in Orion are typical red giant stars. Red giants are mature stars, bloated to enormous size by a host of exotic thermonuclear reactions in and around their cores. Ordinary stars like the sun become red giants when all of the hydrogen at their centers has been consumed.

And there are stellar corpses: white dwarfs, neutron stars, and black holes. These small, compact objects are created from ancient stars that have exhausted all of their sources of thermonuclear fuel.

As you can imagine, stars undergo dramatic changes as they evolve into various kinds of objects over their lifetimes. For example, in about 5 billion years, all of the hydrogen at the sun's center will be used up, and our star will swell to enormous proportions as it becomes a red giant. Still further into the future, the red giant sun will shrivel up into a white dwarf after all of its internal sources of fuel have been exhausted. Stars that are much more massive than the sun end their existence in a spectacular thermonuclear detonation called a *supernova explosion*. During this cataclysm, the star is completely torn apart, and its burned-out core may implode to form a neutron star or black hole.

During the course of its life, a star may develop internal instabilities that cause it to pulsate. The star actually expands and contracts in a regular, periodic fashion. An excellent example of one type of pulsating star is the *cepheid variables*. Like their prototype δ Cephei, all cepheids vary their brightness in a distinctly characteristic fashion. As shown in Figure 2-1, the brightness of a cepheid undergoes a rapid increase followed by a gradual decrease during each cycle. The periods of cepheids (that is, the time it takes to go through one cycle) range from about 2 to 40 days. Polaris, the North Pole Star, is a cepheid variable that has a period of about 4 days.

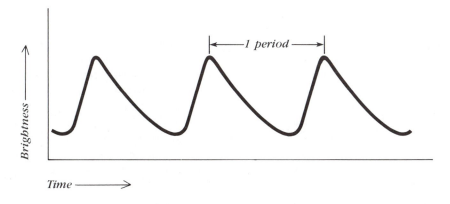

Figure 2-1   A Cepheid Light Curve
A cepheid is a variable star whose brightness changes in a specific fashion: rapid brightening followed by gradual dimming. The most rapid cepheids have periods of only 1 day. Slower cepheids may take slightly more than a month to complete a single cycle.

A cepheid variable is a pulsating star that periodically swells and shrinks. Each time the star expands past its average size, gravity soon stops the expansion, and contraction begins. But like a swinging pendulum, the contracting stellar surface overshoots the equilibrium position and the star becomes somewhat compressed. The buildup of gas pressure inside the compressed star stops the contraction and, like a tightly coiled spring, causes the star's surface to rebound. During each cycle of expansion and contraction, gases in the star's atmosphere are alternately cooled and heated. The characteristic light curve of a cepheid, such as that shown in Figure 2-1, is the combined result of periodic change in the star's surface area and surface temperature.

The most important observations of cepheid variables date back to the pioneering work of Henrietta Leavitt on the Small Magellanic Cloud in 1908. The two Magellanic Clouds are a familiar sight to anyone who lives south of the equator. On a moonless night, they look like two fuzzy patches of light — like two detached pieces of the Milky Way. Through a telescope, they are both resolved into millions of stars. Although Ms. Leavitt could not have known it then, both the Large and Small Magellanic Clouds are nearby galaxies. They are companions of our own Milky Way Galaxy.

While examining data from Harvard College's observatory in the Peruvian mountains, Leavitt discovered an important property of cepheid variables in the Small Magellanic Cloud. She found that dim cepheid variables have short periods, while brighter cepheids have longer periods. But because all of the cepheids in the Small Magellanic Cloud are at roughly the same distance from Earth, Leavitt realized that there must be a direct and simple correlation between the true luminosity of a cepheid and its period. This correlation is today called the *period—luminosity relation.*

Astronomers have put a lot of work into observing cepheids since Leavitt's discovery. For example, in the late 1940s, Walter Baade found that the chemical composition of cepheids affects their brightness. Metal-rich cepheids (that is, cepheids that have substantial traces of heavy elements in their outer gaseous layers) are somewhat brighter than their metal-poor cousins. Incidentally, metal-poor stars are most easily found in globular clusters (see Figure 1-3), while metal-rich stars are found in open clusters (see Figure 1-2). The sun is a typical metal-rich star. Leavitt's work dealt with the metal-rich cepheids. They are usually called *classical cepheids.*

The period—luminosity relation for classical cepheids is shown in Figure 2-3. Notice the clear correlation between brightness and period. For dim cepheids (that is, for cepheids that are only 1,000 times as bright as the sun), the period is only a couple of days. But for bright cepheids (that is, for cepheids that are 20,000 times as bright as the sun), the period is slightly longer than a month.

To the astronomer, cepheid variables are among the most important stars in the sky, all because of the period—luminosity relation. Because of this relation, cepheid variables can be used as "distance

Figure 2-2 The Magellanic Clouds
The Large and Small Magellanic Clouds look like two detached pieces of the Milky Way. They can be seen clearly only from southern latitudes. Actually, both Clouds are nearby galaxies, close enough to us that even small telescopes reveal millions of individual stars. In this photograph, the Large Magellanic Cloud is on the top, and the Small Magellanic Cloud on the bottom. (Yerkes Observatory.)

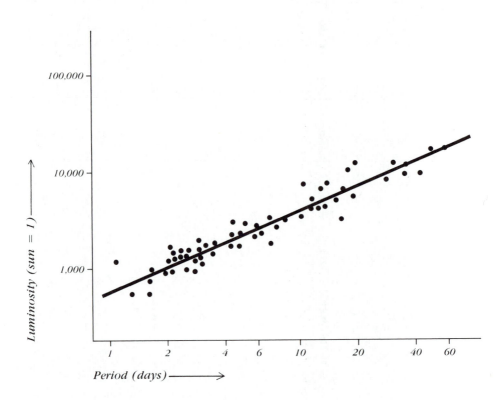

Figure 2-3   The Period–Luminosity Relation
*This graph shows the relationship between the luminosities and
periods of classical cepheid variables. Each dot represents a cepheid
in the Small Magellanic Cloud whose brightness and period have
been measured. The line is the "best fit" to the data. (Adapted from
H. C. Arp.)*

indicators." For example, suppose that you happen to discover a cepheid variable with your telescope. You recognize the star as a cepheid because you have observed it on several nights and have seen the characteristic fashion in which its brightness varies. Using a clock, you measure the cepheid's period. Then you go to the period–luminosity relation (as displayed in Figure 2-3, for example) and from the graph simply read what luminosity goes with your star's period. This gives the *true* brightness of the star. Of course, the cepheid looks a lot dimmer through your telescope. But because you already know the *true* brightness of the star, it is an easy matter to figure out how far away the cepheid must be in order for it to appear as dim as it does. Only one distance is compatible with both the true luminosity and the apparent luminosity of your cepheid.

By the early 1920s, astronomers were divided over the enigma of the "spiral nebulas." Some astronomers felt that these "nebulas" were relatively small and nearby, like all the other objects in Messier's catalog and the NGC. Other astronomers echoed Immanuel Kant's contention that the spiral nebulas are huge whirlpools of stars (Kant coined the phrase "island universes" to describe them) very far from our own Milky Way Galaxy. As I mentioned in Chapter 1, this controversy came to a head in the Shapley-Curtis debate, at which nothing was decided. Nothing was decided because there were no reliable "distance indicators." Nobody had any real understanding of how far away the spiral nebulas are.

While astronomers squared off over the question of spiral nebulas, a young lawyer abandoned the prospects of a lucrative law practice in order to study the stars. He was destined to become one of the greatest astronomers of the twentieth century. His name was Edwin Hubble.

During the night of October 6, 1923, Edwin Hubble took a photograph of the Andromeda "Nebula" with the 100-inch telescope on Mount Wilson. There was a cepheid variable in the field of view. Before long, he had found over a dozen cepheids in the "nebulas" known as NGC 6822 and M33 (see Figure 2-6).

A small portion of the outer edge of the Andromeda "Nebula" is shown in Figure 2-5. Two cepheids are identified. Notice how incredibly faint they look. But we know that cepheid variables are in-

***Figure 2-4 (left)   The Andromeda Galaxy (called M31 or
NGC 224)***
*Hubble's discovery of cepheid variables in this galaxy settled the
Shapley-Curtis debate once and for all. The small white rectangle
shows the field of views of Figure 2-5. (Lick Observatory.)*

Figure 2-5 (above)   Cepheid Variables in M31
*Two cepheids are identified in this view of the outer portions of the
Andromeda "Nebula." Because these stars appear so faint (but
intrinsically they are very luminous), Hubble successfully argued
that they are exceedingly far away. (Hale Observatories.)*

trinsically very luminous. As shown in Figure 2-3, typical cepheids are roughly 10,000 times as luminous as the sun. Hubble immediately realized that the only way these cepheids in M31 could look so dim was if they were far away. Very far away. The distance to M31 turns out to be 2¼ million light years. Only at this enormous distance can those luminous cepheids appear so dim. That places M31 far beyond the outer edges of our own Galaxy.

Hubble's results were presented at a meeting of the American Astronomical Society on December 30, 1924. That settled the Shapley-Curtis debate once and for all. The universe was recognized to be far larger and populated with far bigger objects than anyone had seriously imagined. Hubble had discovered the realm of the galaxies.

There are countless millions of galaxies all across the sky. Galaxies are seen in every unobscured direction. A typical galaxy contains 100 billion stars and measures 100,000 light years in diameter. Galaxies are therefore the biggest individual objects in the universe.

In spite of the vast numbers of galaxies across the sky, Hubble found that they could be classified into one of four broad categories. These categories form the basis for the *Hubble classification scheme.* They are ellipticals, spirals, barred spirals, and irregulars.

We have already seen many fine examples of spiral galaxies. Both M31 and M33 are spirals. While studying spiral galaxies, Hubble found that this class could be further subdivided according to the size of the central bulge and the winding of the spiral arms. Spirals with tightly wound spiral arms and a prominent, fat central bulge are called *Sa galaxies.* Those with moderately wound spiral arms and a moderate-sized central bulge are called *Sb galaxies.* And finally, loosely wound spirals with a tiny central bulge are *Sc galaxies.* The Andromeda galaxy (M31) is a good example of an Sb. The Triangulum galaxy (M33) with its loosely wound arms and tiny nucleus is a typical Sc. You can see M31 and M33 in Figures 2-4 and 2-6, respectively.

Fortunately, the size of the central bulge and the degree of winding of the spiral arms go hand in hand. This permits us to classify spiral galaxies that happen to be viewed nearly edge-on. Thus, for

Figure 2-6   The Triangulum Galaxy (called M33 or NGC 598)
*After several years of careful searching, Hubble discovered 35
cepheids in this galaxy in the constellation of Triangulum. The
distance to this galaxy is 2⅓ million light years. (Lick Observatory.)*

33

Figure 2-7   The Sombrero Galaxy (called M104 or NGC 4594)
*Because of the large size of its central bulge, this galaxy is an Sa. If
we could see it face-on, we would find that the spiral arms are
tightly wound around the voluminous bulge. This galaxy is in the
constellation of Virgo and is tilted by only 6 degrees from our line
of sight. (Kitt Peak National Observatory.)*

example, M104, shown in Figure 2-7, is an Sa galaxy because of its
huge central bulge. In contrast, NGC 4565, shown in Figure 2-8, must
be an Sb galaxy because of its smaller central bulge. The tiny central
bulge of an Sc would hardly be noticeable in an edge-on view.

In addition to ordinary spirals, there are *barred spiral
galaxies.* In barred spirals, the spiral arms originate at the ends of a

Figure 2-8   The Spiral Galaxy NGC 4565
Because of the moderate size of its central bulge, this galaxy must be an Sb. Notice that this central bulge is much smaller than the bulge of M104 in Figure 2-7. NGC 4565 is located in the constellation of Coma Berenices. (Kitt Peak National Observatory.)

bar running through the galaxy's nucleus rather than from the nucleus itself. Figure 2-9 shows a beautiful example of a barred spiral. As in the case of ordinary spirals, Hubble subdivided barred spirals into three classes, depending on the size of the central bulge and the winding of the spiral arms. An *SBa galaxy* has a large central bulge and tightly wound spiral arms. A barred spiral with a moderate bulge and moderately wound spiral arms is an *SBb galaxy.* NGC 1300 in Figure 2-9 is a fine SBb. An *SBc galaxy* has loosely wound spiral arms and a small central bulge.

In addition to the two major types of spirals, a third broad category of galaxies exhibit no spiral structure at all. These are the elliptical galaxies. Hubble chose to subdivide this category according to how round or flattened they look. The roundest elliptical galaxies are called *E0 galaxies.* The flattest elliptical galaxies are called *E7 galaxies.* Elliptical galaxies with intermediate amounts of flattening receive intermediate designations. For example, M49, shown in Figure 2-10, is an E1 galaxy. It is only slightly flattened.

Of course, an E1 or E2 galaxy might actually be a very flattened disk of stars that just happens to be viewed face-on. Or a cigar-shaped E7 galaxy might look spherical when viewed end-on. But we would not know any of these details without a great deal of further observations, such as determining the galaxy's rotation. Fortunately, none of this matters for the Hubble scheme. Hubble chose to classify galaxies only by their appearance to our earth-bound view.

Elliptical galaxies look far less dramatic than their spiral and barred spiral cousins. Ellipticals are virtually devoid of the huge clouds of interstellar dust and glowing gases that give the spirals and barred spirals their striking appearance. Nevertheless, elliptical galaxies come in an enormous range of sizes and masses. Both the biggest and the smallest galaxies in the universe are ellipticals.

M87 in the constellation of Virgo is a fine example of a *giant elliptical.* It is a whopper. It sits at the middle of a large cluster of galaxies and perhaps occasionally gobbles up an innocent spiral that inadvertently wanders too near. We will have more to say about *galactic cannibalism* and the fascinating case of M87 in later chapters.

Whereas giant ellipticals are rather rare, *dwarf ellipticals* are extremely common. Dwarf ellipticals are only a fraction the size of their normal counterparts and may contain only a few million stars. Some nearby dwarf ellipticals contain so few stars that the galaxy is completely transparent. You can actually see straight through the galaxy's nucleus and out the other side.

Figure 2-9   The Barred Spiral Galaxy NGC 1300
*In barred spirals, the spiral arms originate at the ends of a bar
extending through the galaxy's nucleus. NGC 1300 is an SBb galaxy
in the constellation of Eridanus. (Hale Observatories.)*

37

Figure 2-10   The Elliptical Galaxy M49 (also called NGC 4472)
This galaxy in the constellation of Virgo is a typical E1. Elliptical galaxies have almost no dust or gas between the stars and show no evidence of recent star formation. (Kitt Peak National Observatory.)

Figure 2-11   The Giant Elliptical Galaxy M87 (also called NGC 4486)
This huge elliptical galaxy sits at the center of a rich cluster of galaxies in the constellation of Virgo. It is an E0 galaxy and emits large quantities of X rays and radio waves. (Hale Observatories.)

Figure 2-12   The Small Elliptical Galaxy NGC 147
This nearby elliptical galaxy in the constellation of Cassiopeia is an E6 galaxy. Millions of individual stars are seen in this remarkably clear photograph taken through the Palomar 200-inch telescope. Dwarf ellipticals are even smaller and contain far fewer stars than NGC 147. (Hale Observatories.)

Figure 2-13   A Dwarf Galaxy
*A typical dwarf galaxy contains so few stars that you have no
difficulty seeing straight through the galaxy's center. This dwarf
galaxy is located in the constellation of Sextans. (Hale Observatories.)*

Finally, any galaxy that does not conveniently fall into one of Hubble's three broad categories (that is, elliptical, spiral, or barred spiral, as sketched in Figure 2-14) is usually somewhat weird. For one reason or another, these oddballs have a very peculiar or distorted appearance. Hubble lumped all of them together in his fourth category: *irregulars*. The Magellanic Clouds are irregular galaxies.

Although they are in a distinct minority, irregular galaxies are among the most fascinating objects in the universe. As we shall see in later chapters, many irregulars owe their distorted appearance to colossal explosions in their nuclei or to collisions with nearby galaxies.

It was once thought that spirals were the most common kind of galaxy. Indeed, if you look at the galaxies listed in famous catalogs (such as the *Shapley-Ames Catalogue* or the *Reference Catalogue of Bright Galaxies* compiled by the husband-and-wife team of Gerard and Antoinette de Vaucouleurs), you find that nearly three-quarters are spirals. For example, Harlow Shapley listed the proportions as 75 percent spirals (both normal and barred), 20 percent ellipticals, and 5 percent irregulars.

This preponderance of spirals is valid only if you look at just the brightest galaxies in the sky. Today we realize that there are enormous numbers of small and dwarf ellipticals. If we include these underluminous galaxies, the true proportions are 20 percent spirals, 10 percent barred spirals, 60 percent ellipticals, and 10 percent irregulars.

Ancient peoples believed that humanity occupied a special place in the universe. The earth stood immobile at the focus of the cosmos while the sun, moon, planets, and stars revolved daily about our central location. But 400 years ago the earth was dethroned when Nicholas Copernicus successfully argued that we ride upon one of several planets revolving about the sun.

As recently as the 1920s, it was again believed that we occupy a special location. Many astronomers believed that the Milky Way Galaxy was the only sizable object in the universe. Once again we were dethroned when Edwin Hubble proved that our Galaxy is one of millions scattered across the cosmos.

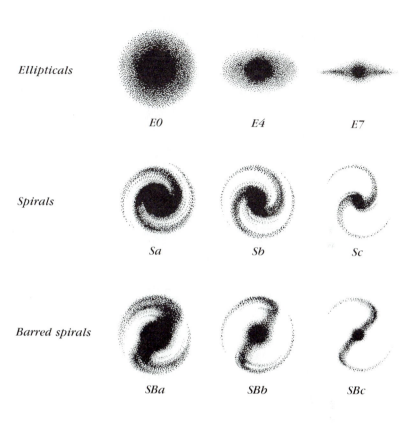

Ellipticals

EO   E4   E7

Spirals

Sa   Sb   Sc

Barred spirals

SBa   SBb   SBc

Figure 2-14 The Hubble Classification Scheme
About 90 percent of all the galaxies in the sky can be classified into one of three broad categories: ellipticals, spirals, or barred spirals. These sketches show further subdivisions of this classification scheme.

43

Figure 2-15   The Large Magellanic Cloud
The Large Magellanic Cloud is a nearby irregular galaxy (although some astronomers see a hint of barred spiral structure). This galaxy is only 160,000 light years away and is a companion of the Milky Way Galaxy. (Cerro Tololo Inter-American Observatory.)

To some people, these astronomical discoveries are depressing. To them, the lesson of modern astronomy is that humanity is a collection of insignificant microbes clinging to a small rock that orbits an ordinary star in an otherwise inconsequential galaxy — just one among billions in an inconceivably vast universe. I prefer a different view. With each new revelation, the human mind must scale new heights and explore new dimensions. Of course, from a purely mechanical aspect, we are a race of Lilliputian life forms precariously poised in the biosphere surrounding a very small planet. But through the human intellect, these tiny creatures have the extraordinary ability to examine and comprehend the structure of the universe. This truly distinguishes us from less-evolved animals. Not how insignificant our bodies are, but rather how potent the human mind is — that is the real lesson of modern astronomy.

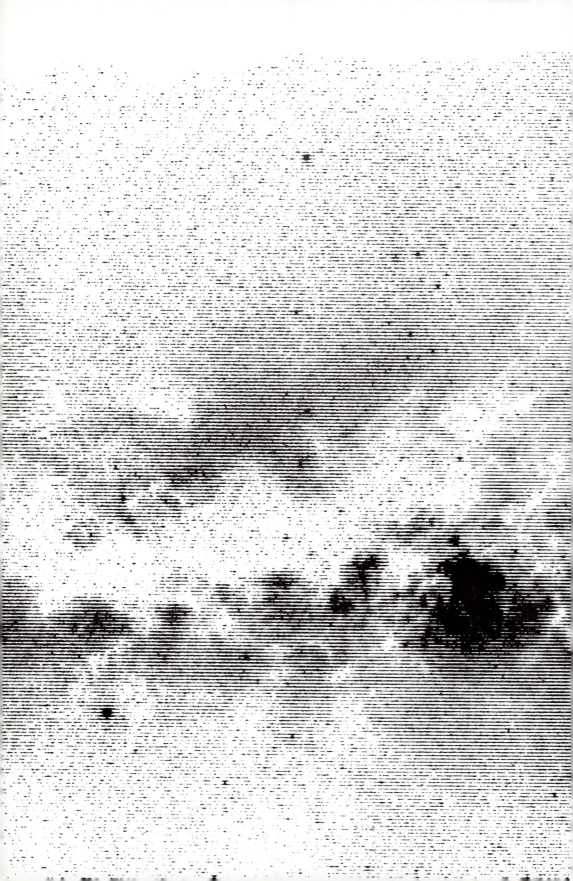

3

Our Galaxy

One of the underlying purposes of astronomy is to clarify our relationship with the rest of the universe. Since the most ancient times, people have turned to astronomers for insight and enlightenment about our location in the cosmic scope of space and time. Of course, only within the last few centuries have the musings of astronomers been backed by telescopic observations.

Sir William Herschel's approach seemed quite straightforward. After all, if you are standing in a forest and you see equal numbers of trees in all directions, then you are probably near the center of the forest. But if instead you see fewer trees in one direction and a higher density of trees in the opposite direction, then you must be located near the edge of the woods. This was the basic idea behind Herschel's attempt to discover our location among all the stars in the Milky Way Galaxy.

Like most astronomers of his day, Herschel realized that the sun is one star among millions in a huge disk-shaped assemblage. By laboriously counting the numbers of stars he could see in 683 selected regions across the sky, Herschel hoped to deduce our location in the Milky Way. His final conclusion was that we are at the center, and his map of the Milky Way Galaxy was shown in Figure 1-8.

Herschel was wrong. Actually, he had no way of knowing that vast quantities of interstellar gas and dust severely limit how far we can see in the plane of the Milky Way. Some of this extensive obscuring material along the Milky Way is readily apparent in Figure 3-1. Herschel had no inkling of the enormous number of stars toward the galactic center, simply because they are hidden from view. It was like trying to deduce your location in a forest filled with mist and fog. Because the mist and fog hide most of the trees, even a careful counting of the numbers of trees in various directions produces an erroneous impression.

Like most of the stars in our galaxy, almost all of the obscuring interstellar gas and dust is concentrated in the plane of the Milky Way. On either side of the Milky Way, we have no trouble seeing distant portions of the universe. For example, in Figure 3-1, the upper third and the lower third of the illustration are almost totally free of any obscuring material. Our first true understanding of the sun's position in the Galaxy was destined to come from observing objects in those unobscured regions on either side of the Milky Way.

Harlow Shapley wanted to be a newspaperman. But upon enrolling at Missouri University, he was disappointed to find that no journalism classes were available. Thumbing through the course catalog, Shapley first came upon several archaeology courses. But he was dissuaded by the unpleasant prospect of a subject filled with archaic and unpronounceable names. So he turned the page to the next entry. At least partly because of the arrangement of the English alphabet, Harlow Shapley became an astronomer.

After receiving his Ph.D. from Princeton University, Shapley moved to California to accept a position at Mount Wilson Observatory. During the next few years, Shapley spent much of his time studying variable stars in globular clusters. As I mentioned in Chapter 1, a globular cluster is a huge, spherical assemblage of stars. A spectacular example is ω Centauri, shown in Figure 3-2, which contains hundreds of thousands of stars.

Globular clusters are made of the oldest stars in the sky. Analysis of the light from globular clusters shows that the outer gaseous layers of these stars are composed almost exclusively of hydrogen and helium. In comparison with stars like the sun, these globular cluster stars contain very small amounts of the heavier elements (this is why they are called "metal-poor" stars, as mentioned in Chapter 2). Hydrogen and helium are the only two elements that could have survived the violent birth of the universe. Thus, all of the first-generation stars that formed from these primordial gases are metal-poor stars. The predominance of metal-poor stars divulges the antiquity of globular clusters.

In contrast to the old stars in globular clusters, the sun is comparatively young. Several generations of rapidly evolving stars had already lived out their lives prior to the sun's birth 5 billion years ago. Many of these ancient, short-lived stars had ended their existences in supernova explosions. During one of these cataclysms, a star is ripped apart, and most of the heavy elements that had been manufactured in the thermonuclear furnaces at the dying star's core

Figure 3-1   The Milky Way
The Milky Way stretches completely around the sky. This hazy band of light is our inside, edge-on view of the galaxy in which we live.

are cast out into space. This process therefore "enriches" the interstellar medium with heavy elements. Later-generation stars that form from this enriched material are therefore destined to be metal rich. A typical metal-rich star like the sun contains (by weight) about 75 percent hydrogen, 20 percent helium, and at most 5 percent for all the remaining heavier elements.

Shapley managed to find variable stars in several globular clusters. Most of these pulsating stars turned out to be *RR Lyrae variables,* a class of variables named after the prototype in the constella-

This meticulous drawing was made from numerous wide-angle photographs. (Lund Observatory.)

tion of Lyra. They are quite similar to the metal-poor cepheids first described by Walter Baade in the 1940s.

RR Lyrae variables, like cepheids, are important because they can be used as distance indicators. All RR Lyrae variables have periods of 10 to 15 hours, and all have nearly the same average true brightness. On the average, they all shine with a luminosity 100 times as bright as the sun. Thus if you find an RR Lyrae variable (which you recognize by its short period and cepheidlike light curve, as shown in Figure 3-3), you can deduce its distance by comparing its apparent

Figure 3-2 The Globular Cluster Omega Centauri (also called NGC 5139)
This globular cluster in the southern constellation of Centaurus is visible to the naked eye as a dim "fuzzy star." Actually, it contains hundreds of thousands of individual stars and is about 17,000 light years away. (Cerro Tololo Inter-American Observatory.)

brightness with its true brightness. For example, Figure 3-4 shows M55, one of the globular clusters that Shapley studied. Three RR Lyrae variables are identified. To find the distance, you simply figure out how far away those stars (which you already know are really 100 times as bright as the sun) must be in order for them to appear as dim as they do in your photograph. The distance to M55 turns out to be 20,000 light years.

By 1915, Shapley had noticed a peculiar property of globular clusters. Ordinary stars and open clusters appear to be rather uniformly spread around the Milky Way. But the majority of the 93 globular clusters that Shapley studied were preferentially located in one part of the sky. The majority are widely scattered around the portion of the Milky Way in Sagittarius.

Within two years, Shapley had mapped out the three-dimensional distribution of the 93 then-known globular clusters. The swarm of globular clusters was centered on a very distant location in the direction of Sagittarius. While puzzling over this clear tendency for globular clusters to be grouped toward one part of the sky, Shapley had a brilliant inspiration: the distribution of globular clusters reveals our true location in the Milky Way Galaxy. The globular clusters seem to be grouped around Sagittarius only because the earth is really located far from the center of the Milky Way. The true center of our Galaxy is the same as the center of the swarm of globular clusters.

In many respects, Shapley's discovery ranks with the revelations of Copernicus four centuries earlier. Copernicus showed that the earth is not at the center of the solar system, and Shapley proved that the sun is not at the center of the Milky Way Galaxy. The galactic center is 30,000 light years away, in the direction of Sagittarius. The entire Galaxy is about 100,000 light years in diameter. Thus, we are located about two-thirds of the way between the center and the edge.

As you can perhaps imagine, optical observations are not much help in determining details of the shape and structure of our Galaxy. Interstellar gas and dust effectively obscure our view. For example, light from stars near the galactic center is dimmed by a factor of 10 billion during the course of its 30,000-year journey to-

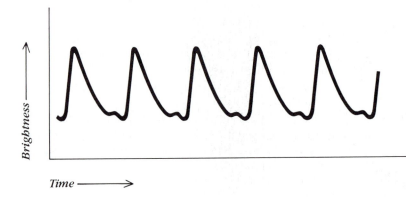

Figure 3-3   An RR Lyrae Light Curve
RR Lyrae variables are recognized by the characteristic way in which they vary their brightness. All RR Lyrae variables have periods of roughly ½ day, and all have an average true luminosity equal to approximately 100 suns.

Figure 3-4   The Globular Cluster M55 (also called NGC 6809)
*Three RR Lyrae variables are identified in this globular cluster
located in the constellation of Sagittarius. By comparing the
average apparent brightness of these stars (as seen in this
photograph) with the average true brightness (known to be roughly
100 suns), astronomers deduce that the distance to this cluster is
20,000 light years. (Harvard Observatory.)*

ward us. We cannot see the galactic center simply because only one ten-billionth of the light can get through all the intervening gas and dust.

Consider a foggy, smoggy day. Although you may not be able to see down the street, your radio and television sets work perfectly. Particles in the air efficiently scatter and absorb visible light, but radio waves from the broadcasting stations stream through unhampered. For this reason, a detailed understanding of the shape of our Galaxy was destined to come from radio astronomy.

Although radio waves from space were discovered in the 1930s, radio astronomy really did not get started until after World War II. At that time, astronomers realized that advances in electronics and electrical engineering during the war had produced much of the technology needed to construct radio telescopes. Radio telescopes began springing up in England, the Netherlands, Australia, and the United States. Because the wavelengths of radio waves are much longer than the wavelengths of visible light, radio telescopes had to be much bigger than their optical counterparts. A fine, modern radio telescope is shown in Figure 3-5.

The advent of radio astronomy was like giving new eyes to a blind person. For thousands of years, all our knowledge about the universe was based entirely on visible light alone. Visible light constitutes only a tiny fraction of a vast spectrum of radiations that could be coming to us from outer space. Radio telescopes gave us the ability to see the universe in a wavelength range far removed from ordinary light. As shown in Figure 3-6, the view is vastly different.

Several natural processes produce radio waves in space. One of these processes is directly related to the structure of the hydrogen atom.

Hydrogen is by far the most abundant substance in the universe. Hydrogen alone accounts for three-quarters of the mass of the stars and gas in our Galaxy.

Hydrogen is the lightest element, and its atoms are the simplest atoms in nature. A hydrogen atom consists of a single electron orbiting a single proton. And like most nuclear particles, electrons and protons are rotating; they have *spin.*

Figure 3-5   A Radio Telescope
The dish of a radio telescope collects and focuses radio waves from space. Electronic equipment is used to amplify and record the signal. This radio telescope is at the National Radio Astronomy Observatory in West Virginia. The diameter of the dish is 140 feet. (N.R.A.O.)

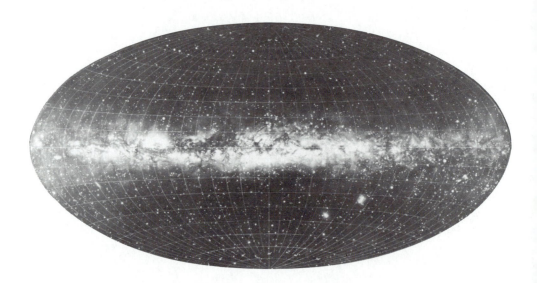

Figure 3-6   The Visible Sky and the Radio Sky
*The drawing at the left shows the entire visible sky. The drawing at
the right shows the entire radio sky. Both illustrations have the
same scale and orientation. Both are centered about the Milky Way.
(Griffith Observatory and Lund Observatory.)*

In the frigid depths of interstellar space, the electrons in hydrogen atoms circle their protons in the lowest possible orbit. In the absence of any external sources of energy (such as a nearby hot star), the electrons remain huddled close to the protons. But according to the laws of quantum mechanics, there are two possible configurations for the electron and the proton in one of these hydrogen atoms. As shown in Figure 3-7, they can be spinning in the same direction or they can be spinning in opposite directions. Furthermore, the electron can flip over. This *spin–flip transition* produces radio waves at exactly 1,420 megacycles, which corresponds to a wavelength of 21.11 centimeters. This radiation is therefore called *21-cm radiation.*

In 1944, even before the end of the war, the Dutch astronomer H. C. van de Hulst predicted that 21-cm radiation would be detected from vast hydrogen clouds in space as soon as decent radio telescopes became available. Over the next seven years, two competing teams of astronomers worked to develop the necessary equipment. The Dutch team in Leiden seemed to be ahead until a fire in 1951 temporarily put them out of business. Meanwhile the American team at Harvard pushed on, but not without some difficulty. Edward M. Purcell and Harold Ewen constructed a "horn antenna," which they suspended outside a window of the physics building at Harvard. They soon found that their funnel-shaped contraption was an inviting target for undergraduates armed with snowballs. It is easier to cope with snowballs than fire, and the Harvard physicists were the first to hear the eerie hiss of cosmic static at 21 centimeters.

The discovery of 21-cm radiation was crucial: it gave us the tools to probe the Galaxy. It was obvious that our Galaxy is some sort of spiral. It is far too flattened to be an elliptical (besides, ellipticals do not have vast ragged lanes of obscuring dust and gas, as seen in Figure 3-1). And the Milky Way is too symmetrical to be an irregular. By mapping the location of 21-cm radiation from hydrogen gas in the spiral arms, astronomers soon began to uncover the structure of our Galaxy. But how could they distinguish 21-cm radiation produced in a nearby spiral arm from the 21-cm radiation emitted from a more distant spiral arm?

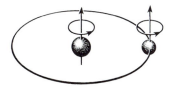
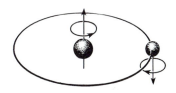

Parallel spins   *Antiparallel spins*

Figure 3-7   Hyperfine Structure of the Hydrogen Atom
In the lowest orbit of the hydrogen atom, the electron can be spinning in either the same direction as or in the opposite direction from the proton. When the electron flips over, the atom either gains or loses a tiny amount of energy. This energy is emitted as radio waves with a wavelength of 21 centimeters.

Imagine standing by the side of a road while a police car or an ambulance races by with its siren wailing. If you listen carefully, you notice that the pitch of the siren is high while the vehicle is approaching you. The pitch drops as the vehicle passes you. This lower pitch is heard as the police car or ambulance recedes from you.

This phenomenon is called the *Doppler effect,* named after the nineteenth-century Austrian physicist Christian Doppler, who first explained it. As shown in Figure 3-8, sound waves in front of the moving vehicle are bunched up. A pedestrian who listens to the approaching siren is exposed to sound waves that have a shorter-than-usual wavelength. And a short wavelength sound is a high-pitched sound. Conversely, as also shown in Figure 3-8, sound waves behind the moving vehicle are spread out. Sound waves from the receding siren, therefore, have a longer-than-usual wavelength, which is equivalent to a low-pitched sound.

This same effect also occurs with light and radio waves. For example, an approaching source of radio waves has its radiations shifted to shorter wavelengths than usual. And conversely, radiations from a receding source are shifted to longer wavelengths than usual. Furthermore, the amount of the wavelength shift is directly proportional to the speed of the source. The higher the speed, the greater the wavelength shift.

All galaxies are rotating. Not only do their beautiful pinwheel shapes clearly suggest rotation, but nothing else would make any sense. If our Galaxy were not rotating, all the stars and gas would fall to the center of the Galaxy, just as all the planets would soon plunge into the sun if they suddenly stopped moving along their orbits. And in addition, different portions of the Galaxy are rotating about the galactic center at different speeds. Just as Mercury goes about the sun faster than Pluto does, stars and gas near the galactic center have a different speed from that of stars and gas farther away.

Suppose that you point your radio telescope toward some direction in the Milky Way. Because radio waves are relatively unhampered by obscuring material, 21-cm radiation from portions of several spiral arms along your line of sight easily reaches your telescope. But different spiral arms are moving at different speeds.

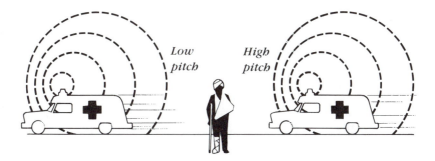

Figure 3-8 The Doppler Effect
*Sound waves in front of an approaching siren are slightly
compressed, so the pedestrian hears a shorter-wavelength, higher-
pitched note than usual. Similarly, sound waves behind a receding siren are slightly spread out, and the pedestrian hears a
longer-wavelength, lower-pitched note than usual.*

Consequently, radiation from hydrogen in these various spiral arms is shifted by various amounts from the usual 21.11-centimeter wavelength.

By way of our example, consider the situation diagrammed in Figure 3-9. Arm *A* is closest to us, and therefore hydrogen clouds in Arm *A* are moving around the Galaxy with a speed nearly equal to that of the sun. Therefore, relative to us, the hydrogen in Arm *A* is hardly moving at all, and its 21-cm radiation has hardly any displacement from the usual wavelength. But in contrast, Arm *C* is much farther away. Gas in this distant arm is going about the Galaxy much more slowly than the sun is. Thus, 21-cm radiation from hydrogen clouds in Arm *C* would be shifted substantially from its usual wavelength. And radiation from Arm *B,* which is at an intermediate distance, would be shifted by an intermediate amount.

The graph in Figure 3-9 shows what the radio astronomer observes. A strong signal with almost no wavelength shift is received from the nearest spiral arm. Weaker signals with larger wavelength shifts are detected from more distant spiral arms. By making extensive observations all along the Milky Way, and by piecing together all the data, radio astronomers have succeeded in mapping large portions of our Galaxy. One of these maps is shown in Figure 3-10. From the winding of the spiral arms, it is clear that we live in an Sb spiral. If we could view our Galaxy from a vantage point several million light years away, it would probably look like M81, shown in Figure 3-11.

Because the 21-cm maps give most of the story, we have no trouble filling in the few gaps. The final, overall picture is shown in Figure 3-12. Notice that our Galaxy really has two components. Of course, the major component is the *disk,* with its beautiful, arching spiral arms. Most of the stars in the disk are the metal-rich variety. But, in addition, our Galaxy has a *halo,* which is defined by the globular clusters. Whereas the disk is very flat, the halo is quite spherical. Most of the stars in the halo are the metal-poor variety.

The sun is located between two major spiral arms. During the summer nights, when we can look toward the galactic center, we see the *Sagittarius Arm.* And during the winter, when our nighttime view is directed away from the galactic center, we see the *Perseus Arm.* Almost every star that you can see with the naked eye is actually

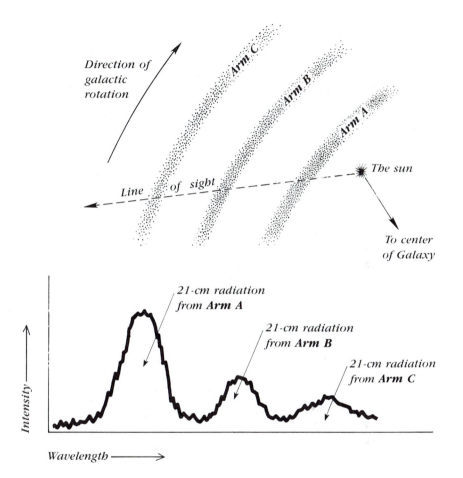

Figure 3-9 Mapping the Spiral Arms
With a radio telescope, 21-cm radiation is detected from hydrogen gas in several spiral arms along the telescope's line of sight. But radiation from each spiral arm is Doppler-shifted according to the relative speed (along the line of sight) between us and that spiral arm. In this way, radio astronomers can sort out signals from different spiral arms and map our Galaxy.

Figure 3-10   A Map of the Galaxy
This drawing was based on radio telescope surveys of 21-cm radiation. The distribution of hydrogen gas across our Galaxy clearly reveals spiral structure. The sun's position is marked by the symbol ⊙. (Courtesy of G. Westerhout.)

Figure 3-11   The Sb Spiral M81 (also called NGC 3031)
Our Galaxy would probably look like M81 if we could see it from
a very great distance. This particular galaxy is about 7 million
light years away, in the constellation of Ursa Major. (Kitt Peak
National Observatory.)

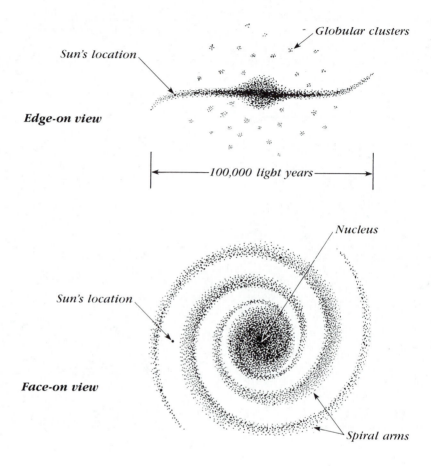

Figure 3-12   Our Galaxy
The sun is located between two spiral arms, as shown in the "face-on view." Our Galaxy's halo of globular clusters is most easily seen in the "edge-on view."

positioned near the sun, in the region between the major spiral arms. All the bright stars that make up the familiar constellations are really huddled quite close to the sun.

It is remarkable to realize that the stars that you see on a clear night are a very tiny fraction of all the stars in our Galaxy. Most of our Galaxy is hidden from view by cosmic gas and dust — the same materials of which we ourselves are composed.

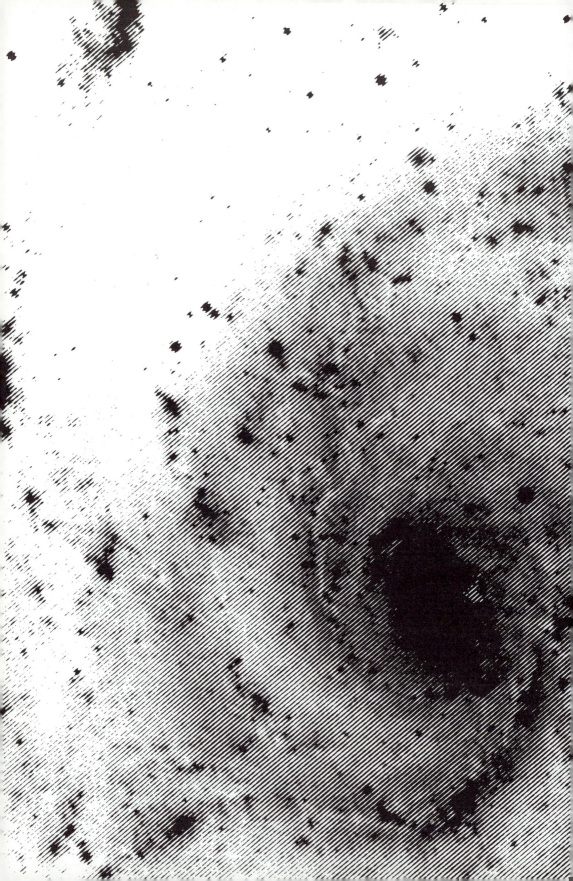

4

The Arms and the Nucleus

Spiral galaxies are certainly among the most beautiful objects in the universe. If only we lived long enough, we could watch as the majestic spiral arms ponderously rotate about the centers of galaxies in a magnificent celestial ballet. But of course it takes too long. The rotation periods of galaxies are measured in hundreds of millions of years. Indeed, a comparison of two photographs of the same galaxy taken several decades apart fails to reveal the slightest hint of movement.

Although we do not actually *see* any motion, we can detect and measure the rotation of galaxies by means of the Doppler effect. As discussed in the previous chapter, the Doppler effect is a shift in wavelength of radiation caused by motion. If a source of radiation is coming toward you, you observe shorter wavelengths than usual. Conversely, if a source of radiation is moving away from you, you see longer wavelengths than usual. The amount of wavelength shift depends on the relative speed (along the line of sight) between you and the source. The higher the speed, the greater the shift.

The rotation of a galaxy can be deduced from measurements of the wavelength shift of the 21-cm radiation from different parts of the galaxy. The approaching side exhibits a shortening of wavelengths, while the receding side displays a lengthening of wavelengths. This same phenomenon also occurs with ordinary visible light. Extensive measurements of the rotation of galaxies based on the Doppler shifts of ordinary light were carried out by Kevin Prendergast along with the husband-and-wife team of Geoffrey and Margaret Burbidge in the early 1960s.

The final result of all these observations of Doppler shifts is a *rotation curve.* As shown in Figure 4-1, a rotation curve is simply a graph of the speed of rotation versus distance from the galaxy's center. The rotation curves of four galaxies are given in Figure 4-1.

The four rotation curves graphed in Figure 4-1 display the same general characteristics shared by all spiral galaxies. In the inner regions of a spiral galaxy (that is, out to roughly 20,000 light years from the center), rotation speed increases with distance. This is typical of *rigid rotation,* such as the rotation of a phonograph record: the farther you are from the axis of rotation, the higher your speed. But in the outer regions of a spiral galaxy, the rotation speed de-

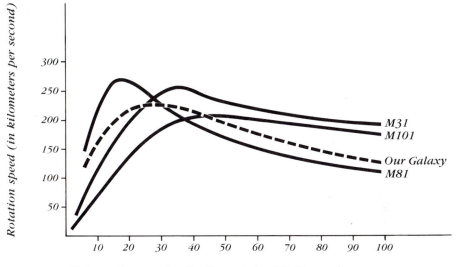

Distance from center (in thousands of light years)

Figure 4-1   Rotation Curves
By measuring the Doppler shifts of light from different parts of a galaxy, it is possible to deduce how that galaxy is rotating. The rotation curves of four galaxies are displayed in this graph.

creases with distance. This is characteristic of *differential rotation,* such as the rotation displayed in the solar system: planets near the sun are moving much faster than planets far from the sun.

At our location 30,000 light years from the galactic nucleus, our speed around the Galaxy is roughly 200 kilometers per second (nearly 500,000 miles per hour). Therefore it takes us 250 million years to complete a single orbit about the galactic nucleus. This period of time is often called the *galactic year.* Because the sun is nearly 5 billion years old, we have been around the Galaxy twenty times.

After a moment's thought about galactic rotation, you perhaps realize that we face a troublesome dilemma. Why are there spiral arms? How can spiral galaxies possibly exist? After all, different parts of a galaxy are rotating at different speeds. We have been around the Galaxy twenty times (and the Galaxy is at least three times as old as the sun). But stars farther from the galactic center are moving much more slowly than the sun. So the spiral arms should wind up. After only a couple of galactic rotations, the spiral arms should become so tightly wound that they disappear. But spiral arms persist. In spite of the differential rotation observed in spiral galaxies, the spiral arms last for dozens of galactic years. How can this be?

The paradox of the spiral arms plagued astronomers for decades. Fortunately, in the late 1960s and early 1970s, the enigma of spiral structure began yielding to the brilliant and ingenious work of several scientists. It all started with the Swedish astronomer Bertil Lindblad, who began struggling over spiral structure in the 1920s — just after Hubble's crucial revelations. Lindblad argued that the spiral arms of a galaxy are merely a *pattern* that moves among the stars. For example, think about waves on the ocean. As the waves move across the surface of the water, the individual water molecules simply bob up and down in little circles. A cork in the water simply bobs up and down as the waves ripple by. The waves are simply a pattern that moves across the water. Indeed, Lindblad spoke of *density waves* in discussing the possible cause of spiral structure.

This density wave theory was greatly elaborated and mathematically embellished by the American astronomers C. C. Lin and Frank Shu in the mid-1960s. Lin and Shu argued that density

Figure 4-2   The Spiral Galaxy M101 (also called NGC 5457)
Different parts of a galaxy rotate at different speeds. Spiral arms should therefore wind up completely after only a few galactic rotations. The obvious persistence of spiral arms has been one of the great mysteries of modern astronomy. (Kitt Peak National Observatory.)

waves passing through the disk of a galaxy cause material to pile up temporarily. A spiral arm is therefore simply a temporary enhancement or compression of the material in a galaxy.

Imagine some workers painting a line down a busy freeway. Normally the cars cruise down the freeway at 55 mph. But because of the crew of painters, there is a temporary bottleneck. The cars must slow down temporarily to avoid hitting anyone. As seen from the air, there is a noticeable congestion of cars around the painters. An individual car spends only a few moments in the traffic jam before resuming the usual 55 mph speed. But the traffic jam lasts all day long, inching its way down the street. The traffic jam — which would be seen so clearly from an airplane — is simply a temporary enhancement of the number of cars in a particular location.

But what is a density wave? What happens in a galaxy that would be like a crew of workers on a freeway? Although we are still far from a complete understanding, important progress was made in 1973 by Agris J. Kalnajs. We must begin by studying the motions of stars in a galaxy.

Think about the stars in a galaxy. Although things might look a little congested in a photograph, there really is a lot of space between the stars. In fact, typical separations between stars are so large (after all, look around at the nighttime sky!) that collisions between stars almost never happen. Instead, all the stars simply orbit the galaxy along nearly circular orbits, just as the planets orbit the sun.

Once again, think about a water wave on the ocean. *Any* wave is a departure from equilibrium. In the case of a water wave, a disturbance starts the water molecules pushing against each other. One molecule pushes the next, which pushes the next, and so on. An individual molecule rotates around in a tiny ellipse as the wave pattern moves across the water, as shown in Figure 4-3.

In a galaxy, the stars are separated by vast distances. But the stars do interact because of gravity. Stars feel each other's gravitational fields. In water waves or sound waves, molecular forces are responsible for orchestrating the motions of molecules. In a galaxy, the force of gravity controls the interactions between stars.

If undisturbed, a star orbits the center of a galaxy along a nearly circular path. But if there is some sort of disturbance, the star

76

Water wave

Kinematic wave

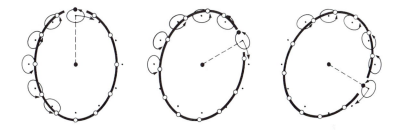

Figure 4-3   Water Waves and Kinematic Waves
*In a water wave, the molecules rotate about the undisturbed water
level in tiny ellipses. Similarly, a small disturbance in the orbit of a
star can cause the star to oscillate in tiny ellipses about its original
orbit. As demonstrated in the lower half of this illustration, the
final path of the star is a precessing ellipse. (Adapted from
A. Toomre.)*

is pushed slightly out of its equilibrium orbit. Just as a water
molecule bobs up and down on the surface of the ocean, the star will
oscillate back and forth about its original orbit. The final result, called
a *kinematic wave,* is shown in Figure 4-3. Notice that the final orbit
of the star is a *precessing ellipse,* an ellipse that rotates. Even though
an individual star simply oscillates about a tiny ellipse that moves
along the star's original circular orbit, the real path of the star (obtained by simply connecting the dots, as in a child's puzzle) is a precessing ellipse. Of course, the gravity of this star affects the motions
of its neighbors, and the original wave disturbance propagates from
one stellar orbit to the next.

Kalnajs had the important realization that either the orbits of stars are organized or they are not. If the orbits are completely random, then the situation looks like the upper half of Figure 4-4. But suppose that the orbits of stars are *not* randomly oriented. Suppose that there is a strict correlation between orbits: each precessing elliptical orbit is tilted with respect to its neighbor through a specific angle. Then the situation looks like the lower half of Figure 4-4. Notice how the beautiful spiral pattern emerges!

Several of Kalnajs's spiral shapes are shown in Figure 4-5. Notice that the spiral pattern arises in those locations where the ellipses are bunched closest together. Of course, stars in a galaxy are scattered all over the place. But with correlated orbits (as in Figure 4-5), some of the stars occasionally get close together along huge arching spirals. After all, that is where the stars' orbits are closest together for the longest stretches of distance.

The temporary enhancement of stars has a profound effect on the interstellar gas and dust. Because of the presence of extra stars, there is an increased gravitational attraction all along the spiral. This has almost no effect on the massive stars that simply continue to lumber along their orbits. But the lightweight atoms and molecules in the interstellar medium are readily sucked into the gravitational well along the spiral. This forms the crest of the density wave.

As Kalnajs's spiral patterns precess, the density waves move through the material of a galaxy at a speed of roughly 30 kilometers per second. But on its own, the interstellar gas can transport a disturbance (such as a slight compression) at a speed of only 10 kilometers per second, which is the speed of sound in the interstellar medium. Thus, the density wave is *supersonic,* because its speed through the interstellar gas is three times the speed of sound in that gas. And just like the case of a supersonic airplane traveling through the air, a *shock wave* is created all along the leading edge of the density wave. Shock waves are characterized by a sudden and abrupt compression of the medium through which they move (you hear a "sonic boom" from supersonic airplanes). The interstellar medium is violently compressed by the cosmic sonic boom of the density wave.

This density wave theory seems to explain many of the properties of spiral structure. For example, take a close look at the spiral

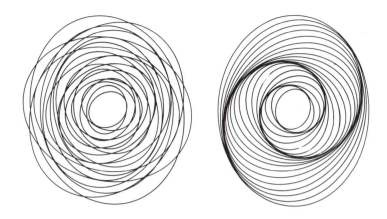

Figure 4-4   The Origin of Spiral Density Waves
*Both drawings have exactly the same number of ellipses, each
ellipse representing the orbit of a star. In the drawing at the
left, ellipses are randomly oriented. In the drawing at the right,
there is a correlation between adjacent ellipses.*

arm shown in Figure 4-6. All those bright spots along the spiral arm
that look like stars are not stars. The galaxy is so far away that it is
almost impossible to see any individual stars. Instead, those bright
features outlining the spiral arms are *emission nebulas*. The Tarantula Nebula shown in Figure 4-7 and the Orion Nebula in Figure 1-4
are good examples of emission nebulas.

An emission nebula is a huge cloud of gas studded with a mul-
titude of very young stars. Many of these recently created stars
radiate large amounts of ultraviolet light. This intense, energetic radiation easily excites the surrounding hydrogen gas, causing it to glow.
In the Tarantula Nebula, a central cluster of newborn blue supergiant
stars produces enough ultraviolet radiation to excite hydrogen gas
out to a distance of 400 light years. This vast, fluorescing cloud
shines with a luminosity equal to half a million suns.

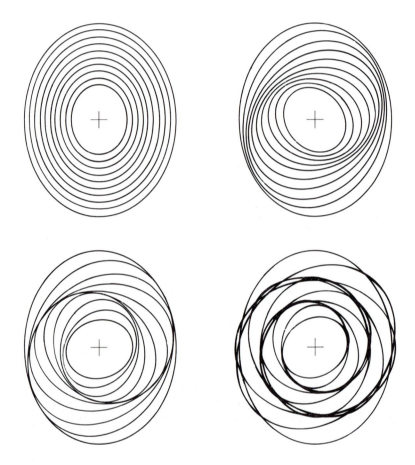

Figure 4-5 Spiral Patterns
Different spiral patterns can be obtained simply by superimposing ellipses. This suggests that spiral density waves are created by the superimposition of elliptical kinematic waves of the type shown in Figure 4-3. (Adapted from A. Kalnajs.)

Figure 4-6   Details of a Spiral Arm
*Spiral arms are outlined by huge glowing clouds of gas called
emission nebulas. Also notice that prominent, dark dust lanes skirt
the inner (concave) edge of the spiral arm. (Hale Observatories.)*

81

Figure 4-7   The Tarantula Nebula (also called 30 Doradus or NGC 2070)
This beautiful glowing cloud of gas is the brightest emission nebula in the Large Magellanic Cloud. This nebula is nearly 30 times as large as the Orion Nebula (see Figure 1-4) and shines with a brightness equaling half a million suns. (Cerro Tololo Inter-American Observatory.)

When a density wave sweeps through the interstellar medium, the gas and dust are suddenly compressed. This compression triggers star formation. Knots, condensations, and globules in the compressed material can contract under the influence of their own self-gravity. Eventually, temperatures and densities deep inside these contracting globules get so high that thermonuclear fires are ignited. Stars are born. The largest and brightest of these newborn stars emit enough ultraviolet light to cause all the surrounding gases to fluoresce. An emission nebula blazes forth. Numerous contracting globules and newborn stars are scattered throughout the Lagoon Nebula, shown in Figure 4-8.

As the spiral density waves sweep through the plane of a galaxy, they recycle the interstellar medium. Old gas and dust that were left behind from ancient, dead stars are compressed into new nebulas in which new stars are formed. As first noted by M. Fujimoto and W. W. Roberts in the late 1960s, the sprawling dust lanes alongside the string of emission nebulas that outline a spiral arm attest to the recent passage of a compressional shock wave. Because the material left over from the deaths of ancient stars is enriched in heavy elements, new generations of stars are more metal rich than their ancestors. The overall structure of spiral shocks and density waves in our own Galaxy is shown schematically in Figure 4-9.

You would be sorely misled to think that we have solved all the questions about spiral structure in galaxies. I think it is now clear that astronomers are on the right track, but there are still many gaps and loopholes in our understanding. For example, what in the world keeps the density waves going? Why don't the density waves simply fade away? Density waves expend an enormous amount of energy to compress the interstellar gas and dust. In order to keep the density waves going, there must be a constant replenishing of that energy. We really do not understand where this energy comes from. But the nuclei of galaxies seem to be the obvious place to look.

The nucleus of our Galaxy is a very active, crowded place. The stars in Figure 4-10 give some hint of the stellar congestion. If you lived on a planet near the galactic center, you would see a million stars as bright as Sirius, the brightest single star in our own nighttime sky. The total intensity of starlight from all those nearby

Figure 4-8   The Lagoon Nebula (also called M8 or NGC 6523)
*This beautiful emission nebula is about 6,500 light years from
Earth, in the constellation of Sagittarius. Many dark globules are
seen silhouetted against the bright background. This nebula is a
site of active star formation. The diameter of the nebula is about 60
light years. (Kitt Peak National Observatory.)*

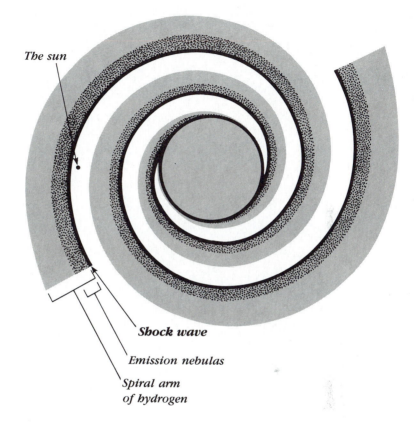

Figure 4-9   *Shock Fronts and Density Waves in Our Galaxy*
A shock wave accompanies the spiral density wave as it sweeps around our Galaxy. The sun is located 30,000 light years from the galactic nucleus.

Figure 4-10   A View Toward the Galactic Center
*More than a million stars are seen in this photograph. This view
looks toward a relatively clear "window" just 4 degrees south of the
galactic nucleus. There is surprisingly little obscuring matter in this
tiny section of the sky. The two globular clusters are NGC 6522 and
NGC 6528. (Kitt Peak National Observatory.)*

stars would be equivalent to 200 of our full moons. Night never really falls on a planet near the center of our Galaxy.

As you might expect, some of our most important information about the galactic nucleus comes from radio observations rather than from visible light. The pioneering observations were first made by Jan H. Oort and G. W. Rougoor in 1960. By observing Doppler shifts of 21-cm radiation, Oort and Rougoor discovered two enormous expanding arms of hydrogen gas. One arm is located between us and the galactic center and is approaching us at a speed of 53 kilometers per second. The other arm is on the other side of the galactic nucleus and is receding from us with a speed of 135 kilometers per second. The total amount of hydrogen in these expanding arms is at least several million solar masses. Something quite extraordinary must have happened about 10 million years ago in order to expel such an enormous amount of gas from the center of our Galaxy.

In addition to the 21-cm radiation from the expanding arms, radio astronomers also detect a vast amount of radio noise coming directly from the galactic nucleus. But the radio noise does not come from hydrogen. Instead it is produced by high-speed electrons spiraling around a magnetic field. This kind of radio emission is called *synchrotron radiation,* and the powerful source at the galactic center is named *Sagittarius A.* Sagittarius A is one of the brightest radio sources in the entire sky. Recent observations suggest that Sagittarius A is very small. In spite of its enormous energy output, Sagittarius A is only 40 light years in diameter.

Three different views of the center of our Galaxy are shown in Figures 4-11a, b, and c. Figure 4-11a is simply a wide-angle photograph showing the Milky Way in the constellation of Sagittarius. The small white rectangle on this photograph gives the field covered by the radio map in Figure 4-11b. Radio astronomers often display their observations in a contour map, each contour line representing a certain level of radio intensity. The small square gives the field of Figure 4-11c, which covers the core of Sagittarius A. As shown in Figure 4-11c, several extremely bright infrared sources are clustered around the galactic center. Each of these sources emits nearly a million times as much energy in infrared radiation as the sun emits at *all* wavelengths combined.

a

Figure 4-11a   The Galactic Center at Optical Wavelengths
*This wide-angle photograph is centered about the Milky Way in
Sagittarius. The small white rectangle shows the region covered by
the radio map in Figure 4-11b. (Hale Observatories.)*

Figure 4-11b   The Galactic Center at Radio Wavelengths
*This radio map shows details of the galactic center at a wavelength
of 3.75 cm. The small black square shows the region covered by the
radio map in Figure 4-11c. (Adapted from D. Downes and
A. Maxwell.)*

Figure 4-11c   Details of Sagittarius A
*This radio map of Sagittarius A was made by Bruce Balik at the
National Radio Astronomy Observatory. The five stars indicate the
locations of five intense sources of infrared radiation discovered by
Frank J. Low and George H. Rieke. (Adapted from R. H. Sanders and
G. T. Wrixon.)*

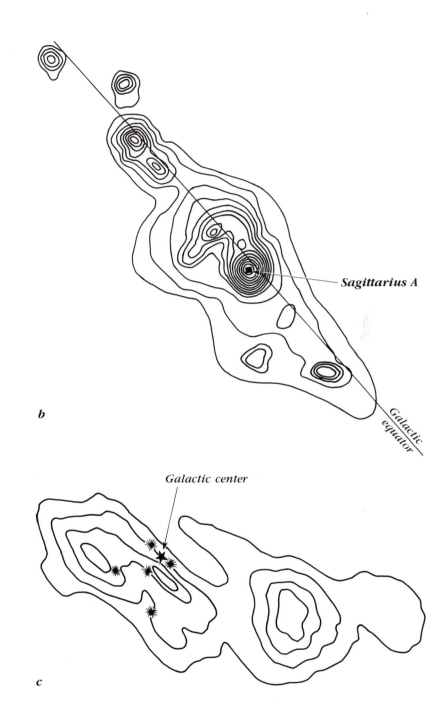

Sagittarius A

Galactic equator

b

Galactic center

c

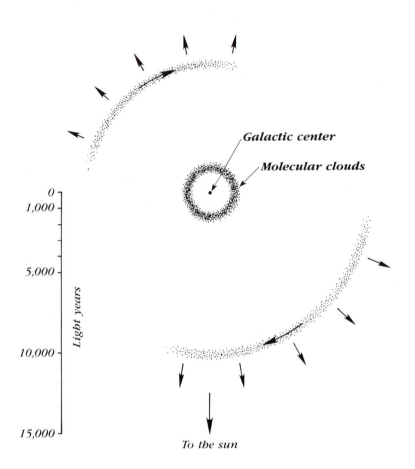

Figure 4-12   The Center of Our Galaxy
*The galactic nucleus is surrounded by a ring of molecular clouds.
Two large expanding arms of hydrogen gas are located farther from
the center. This outward motion suggests that an explosive event
recently occurred at the center of our Galaxy.*

In addition to 21-cm radiation from hydrogen clouds and synchrotron radiation from Sagittarius A, there is a third powerful source of radio waves at the galactic center. Vibrating molecules in huge molecular clouds surrounding the galactic nucleus are strong sources of microwaves. By studying microwaves from the galactic center, astronomers have succeeded in identifying a wide range of chemicals that must be present in these molecular clouds. These substances include ammonia (NH_3), formaldehyde (H_2CO), methanimine (H_2CNH), cyanoacetylene (HC_3N), methyl cyanide (CH_3CN), and many, many others. These molecules are evidently clustered in a ring of huge clouds that encircle the galactic nucleus. One of the largest clouds in this ring is called Sagittarius B2; it contains over a million solar masses of gas and dust. The ring is about 2,000 light years in diameter and is expanding at a speed of 40 kilometers per second. Once again there seems to be clear evidence of a recent explosive event at the center of our Galaxy.

Primarily from radio observations, we have been able to assemble a picture of the central regions of the Galaxy. The basic features are shown in Figure 4-12. As you can perhaps imagine, an enormous amount of energy is needed to account for the outward motions of the two hydrogen arms and the ring of molecular clouds. Traditional concepts in astronomy are at a loss to explain the required amount of energy. Astronomers have therefore begun to examine some exotic ideas. To quote Jan Oort, "Something out of the ordinary appears to be required." The most promising ideas involve a *supermassive black hole.*

As we shall see in a later chapter, the explosive event at the center of our Galaxy is quite tame and mild compared with the violent cataclysms that are tearing apart other galaxies. The enormous energy associated with the intense gravitational fields of supermassive black holes seems to be the only clear hope of accounting for these cosmic catastrophes.

5

The Redshift and the Universe

Ever since the time of Isaac Newton, people have argued that the universe is infinite. Sometimes these arguments were based on mathematical or theoretical considerations. More recently, there has been observational evidence suggesting that the universe is indeed limitless. Let us suppose that this is correct.

What does it mean to say that we live in an infinite universe? First of all, it means that the universe has no edge. There is no boundary, no limit to the size of the universe. Space simply extends forever. And, of course, the universe has no center. If the universe extends infinitely far in all directions, there is no location that could justifiably be called the "center."

One of the great lessons of modern science is that there is nothing particularly unique about our location in space and time. We orbit an ordinary star in an ordinary galaxy — one of billions scattered across the cosmos. This belief is strongly supported by observational evidence. Of course, the details of what we see in the sky vary from one place to another. But on the largest scale, on the scale of millions of light years, the universe is remarkably uniform and homogeneous.

This belief is called the *cosmological principle*. As far as we can tell, the large-scale properties of the universe are the same everywhere. Although there are local variations, all people on all planets see the same overall, large-scale picture of the universe. Every creature on every planet sees galaxies and clusters of galaxies (just as we do) spread across the depths of space.

If the universe is infinite, and if the large-scale properties of the universe are the same everywhere, then there must be an infinite number of stars and galaxies in the universe. This apparently trivial statement has a profound ramification. If there is an infinite number of stars and galaxies in our infinite universe, then you will see a star in *every* unobscured direction in space. Regardless of the direction toward which you point your telescope, your line of sight *must* eventually strike the surface of a star.

So why is the night sky dark? The entire night sky should be dazzlingly brilliant, because you should see the hot, bright surface of a star in every direction. Of course, foreground objects such as tall trees, the moon, clouds, and gas and dust in the Milky Way do get in the way. But on the average, all those extensive unobscured regions

Figure 5-1   A Cluster of Galaxies
Galaxies are seen in every unobscured portion of the sky. This photograph shows a remote but rich cluster of galaxies in the constellation of Coma Berenices. More than a thousand galaxies are seen in this view. This cluster is 440 million light years from Earth. (Hale Observatories.)

of the sky should be as bright as the surface of the sun. How can it be that it is dark during the nighttime?

The dilemma of the dark night sky is called *Oblers paradox,* named after the nineteenth-century astronomer Heinrich Oblers, who first drew attention to the quandary. Of course, we could make some ad hoc assumptions that would dispose of the paradox. But these proposals would conflict with observations of the extent and homogeneity of the universe. Instead, the true resolution of the paradox was destined to involve the evolution of the universe as a whole.

In 1845, the Dutch meteorologist Christopher Heinrich Dietrich Buys-Ballot performed an interesting experiment on a railroad track near Utrecht. He persuaded an orchestra of trumpeters to ride on a flatcar while playing the same note with all their might (to overpower the noise from the locomotive). By listening carefully to the apparent change in pitch as the trumpeters whizzed by, Mr. Buys-Ballot convinced himself of the validity of a phenomenon described three years earlier by Christian Doppler, a professor of mathematics in Prague. As mentioned in Chapter 3, this phenomenon is called the *Doppler effect.*

This same phenomenon also occurs with light. As diagrammed in Figure 5-2, wavelengths from an approaching source of light are shortened. Conversely, wavelengths from a receding source of light are lengthened. In order to measure and record these wavelength shifts, scientists use a *spectrograph.* As diagrammed in Figure 5-3, a spectrograph consists essentially of a prism that breaks up the incoming white light into the colors of the rainbow. Careful examination of this rainbow or *spectrum* often reveals a series of dark lines among the colors. These dark lines are caused by cool gases surrounding the source of light. As light passes through these gases, atoms in the gas selectively absorb some of this radiation at certain specific wavelengths. Each chemical element has its own characteristic pattern of *spectral lines.* The wavelengths of the spectral lines of various chemicals have been determined in laboratory experiments and are listed in standard reference books.

Spectral lines in the light from stars and galaxies are the indelible wavelength-markers by which astronomers can measure

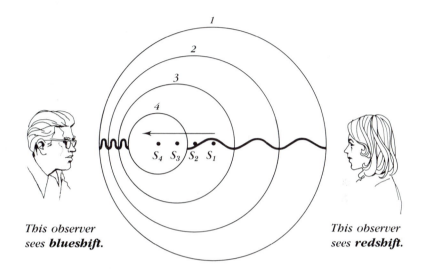

This observer
sees **blueshift**.

This observer
sees **redshift**.

Figure 5-2   The Doppler Effect
*Radiation from an approaching source of light is compressed to
shorter wavelengths than usual. Radiation from a receding source
of light is stretched to longer wavelengths than usual. The amount
of wavelength shift depends on the speed of the object as it moves
between the source and the observer. The higher the speed, the
greater the shift.*

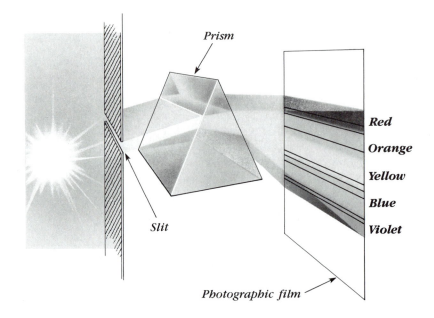

Figure 5-3   A Spectrum and Spectral Lines
When light from a star or galaxy is passed through a prism, the light is dispersed into the colors of the rainbow. This spectrum often contains dark spectral lines that were produced by atoms in the star or galaxy. The spectrograph that an astronomer uses at a telescope also has several lenses to magnify and focus the spectrum onto a photographic plate.

Doppler shifts. Indeed, a spectral line is like one of Buys-Ballot's trumpeters.

Suppose an astronomer photographs the spectrum of a galaxy by mounting a spectrograph at the focus of a telescope. While examining the photographic plate on which the spectrum is recorded, the astronomer recognizes a set of spectral lines as caused by a particular well-known chemical. The astronomer then simply consults a reference book to determine what the wavelengths of these lines should be for a source of light that is not moving. The spectral lines in the spectrum of a galaxy are usually shifted substantially away from their "rest wavelengths." By measuring the size of the wavelength shift, the astronomer easily deduces the speed of the galaxy.

Vesto Slipher of Lowell Observatory was the first astronomer to observe and examine the spectra of galaxies. Between 1912 and 1925, he photographed the spectra of 40 galaxies. All but two displayed Doppler shifts toward longer-than-usual wavelengths. This meant that the overwhelming majority (38 out of 40) of these "spiral nebulas" are rushing away from us. And the measured speeds were surprisingly high. Slipher discovered speeds up to 5,700 kilometers per second (13 million miles per hour, 2 percent of the speed of light). That was a huge velocity for those days. But back then, nobody knew what a "spiral nebula" was anyway. So Slipher's work simply made the whole business more mysterious than ever.

A Doppler shift toward shorter-than-usual wavelengths means that a source of light is coming toward you. Of all the colors of the rainbow, blue light has the shortest wavelength. Thus, astronomers speak of the *blueshift* of light from an approaching star or galaxy. Conversely, a Doppler shift toward longer-than-usual wavelengths means that a source of light is moving away from you. Because red light has the longest wavelength of all the colors of the rainbow, the Doppler shift of a receding star or galaxy is usually called a *redshift.*

Slipher was the first astronomer to notice the clear predominance of redshifts over blueshifts in the spectra of galaxies. In fact, almost all galaxies (except for the nearest) exhibit substantial redshifts in their spectra. Astronomers usually denote the size of the redshift by the letter z. It is simply the fractional shift in wavelength as measured on the spectroscopic plate. For low velocities, z is equal to the speed of the galaxy expressed as a fraction of the speed of light. Thus, for example, a galaxy whose spectrum shows a 12 percent shift in wavelength is therefore receding from us at 12 percent of the speed of light and is said to have a redshift of $z = 0.12$.

For speeds greater than about a third of the speed of light, the simple relation between z and velocity must be modified according to the Special Theory of Relativity. Although it is possible to observe very high redshifts, it is not possible to go faster than the speed of light. Thus, for example, if an object is receding from us so rapidly that its wavelengths are lengthened by 200 percent, then $z = 2$. But this does not mean that the object is traveling at twice the speed of light. Instead, according to the Special Theory of Relativity, a redshift of $z = 2$ corresponds to a speed of 80 percent of the speed of light. The relationship between z and speed is shown graphically in Figure 5-4.

The significance of Slipher's discovery that "spiral nebulas" exhibit a preponderance of redshifts could not be immediately appreciated, because there was no consensus about the locations of these puzzling objects. But as soon as Edwin Hubble had proved that the "spiral nebulas" are in fact very far away, things began to fall into place quite rapidly. In fact, during the late 1920s, Hubble noticed that nearby galaxies have small redshifts, whereas more distant galaxies have significantly larger redshifts. This tendency is clearly displayed in Figure 5-5. The photographs of five elliptical galaxies are shown, arranged in order of increasing distance. To the right of each photograph is the galaxy's spectrum. Two spectral lines (the H and K lines of calcium) are seen in each spectrum. For the nearest galaxy (which looks big simply because it is relatively nearby), the H and K lines are almost exactly where they should be — among the blue colors of the rainbow. But as we turn our attention to more distant galaxies (which look smaller because they are farther away), we notice that the H and K lines are shifted more and more toward the other side of the spectrum. In fact, the most distant galaxy shown in Figure 5-5 has its spectral lines shifted all the way from the blue side of the spectrum to the red side.

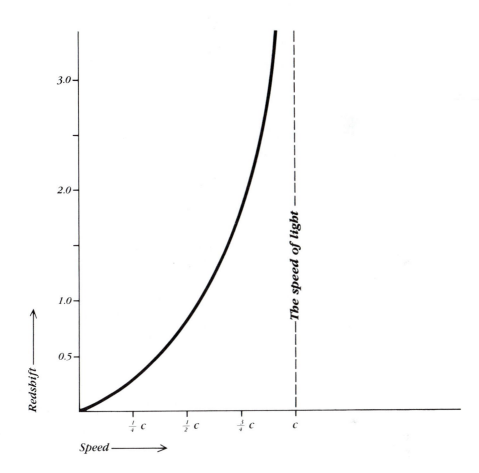

Figure 5-4   Redshift Versus Speed
The redshift z is the fractional shift in wavelength, as measured on a spectroscopic plate. For example, if all the wavelengths of the spectral lines of a galaxy are lengthened by 25 percent, then z = 0.25. This graph shows the relationship between z and speed. The highest possible speed is c, the speed of light.

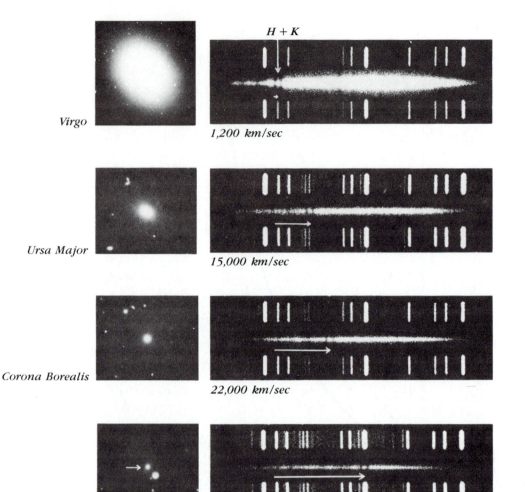

H + K

Virgo

1,200 km/sec

Ursa Major

15,000 km/sec

Corona Borealis

22,000 km/sec

Boötes

39,000 km/sec

Hydra

61,000 km/sec

Figure 5-5   Galaxies and Their Spectra
*Five galaxies (labeled according to the constellation in which they
are located) are shown along with their spectra. The nearest
galaxies have the smallest redshifts, while the more distant galaxies
have larger redshifts. (Hale Observatories.)*

By 1929, Hubble had accumulated enough data to show that there is a direct proportionality between distance and redshift for galaxies. Because of its implications, this relationship ranks as one of the most important scientific discoveries of the twentieth century. The relationship is called the *Hubble law*.

The Hubble law is shown graphically in Figure 5-6. Notice that the Hubble law is a simple linear correlation between distance and speed. Nearby galaxies are moving away from us slowly, whereas more distant galaxies are rushing away from us more rapidly. Double the distance to a galaxy and its speed doubles. Triple the distance and the speed triples.

In order to appreciate the significance of the Hubble law, it is useful to turn to such an analogy as the often-cited *raisin cake analogy*. Imagine making a raisin cake by mixing some dough, raisins, and yeast. You form the loaf and set it on a table, as shown in the "before" view of Figure 5-7. Pick any raisin. That raisin is *your* raisin. Measure the distance between your raisin and several other raisins in the cake. Soon the yeast causes the dough to rise. Suppose that after an hour the raisin cake has doubled in size. Once again you measure the distances between your raisin and other raisins. All the distances are twice as large as before, because the entire cake has doubled in size. And by comparing the changes in distance during the elapsed hour, you discover that nearby raisins are moving away from your raisin rather slowly. But more distant raisins are rushing away much more rapidly. In fact, you discover a direct linear relation between distance and speed. The reason for this linear relation is that the raisin cake is expanding. By analogy, the true meaning of the Hubble law is that *the universe is expanding*.

Of course, this raisin cake analogy has some shortcomings. The cake has an edge and a center, but the universe does not (so, try to imagine a raisin cake that is infinitely huge). Nevertheless, many important properties of the expanding universe are contained in the analogy. For example, our little story does not depend on which raisin you choose as "your raisin." As seen from *any* raisin, all the other raisins are receding with a speed that is proportional to distance. This is in total agreement with the cosmological principle. Any

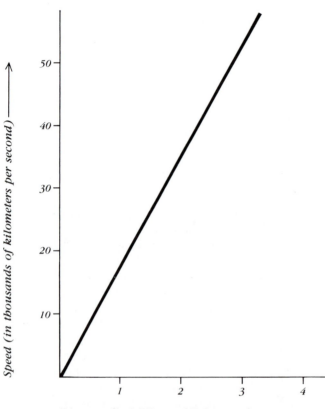

Figure 5-6   The Hubble Law
*The Hubble law is the direct proportionality between the speeds
and distances of galaxies. Nearby galaxies are moving away from
us slowly. More distant galaxies are rushing away from us with
correspondingly higher speeds.*

"Before"   "After"

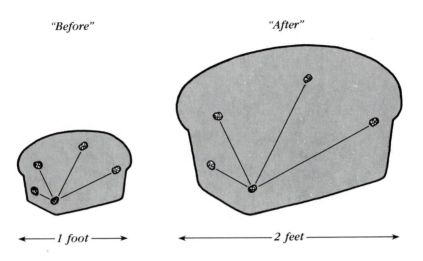

← 1 foot →   ← 2 feet →

Figure 5-7   The Raisin Cake Analogy
*By examining the motions of raisins in an expanding raisin cake,
you discover that a Hubble-law-like relationship exists between
distance and speed. Closely spaced raisins are moving away from
each other rather slowly. Remotely spaced raisins are rushing away
from each other with much higher speeds.*

105

creature on any planet in any galaxy will see all the other distant galaxies receding with speeds that are proportional to their distances.

This realization is enormously gratifying. Just because we see all the galaxies rushing away from us does *not* mean that we are at the "center" of the universe. Very much to the contrary, it means that we are at an entirely typical location in a homogeneous universe. Astronomy books written by intelligent creatures millions of light years away should have a graph that looks essentially the same as Figure 5-6.

The discovery of the expansion of the universe also resolves Oblers paradox. As we gaze farther and farther outward toward distant galaxies, we find that their light is increasingly redshifted. Before long, we find that their light is shifted from visible wavelengths into invisible infrared and radio wavelengths. Although our line of sight does eventually intersect the surface of some remote, dazzlingly bright star, its light is severely redshifted to invisible wavelengths. The sky is dark at night because we live in an expanding universe!

When we look at a distant galaxy, we see its light redshifted because the galaxy is receding from us — perhaps at a sizable fraction of the speed of light. But creatures living on a planet in that distant galaxy do not notice anything unusual. Everything looks quiet and peaceful; they aren't going anywhere. Instead, as they observe *our* Galaxy, they discover that we are rushing away from them at some enormous speed. They find that their stars, candles, and flashlights are emitting radiation at the usual unshifted wavelengths. But they observe that light from our Galaxy has suffered a substantial redshift.

This little story follows directly from the cosmological principle. All it says is that, from the viewpoint of the physical laws that govern the universe, our planet is not any better than anyone else's. But it also suggests an improved way of thinking about the redshift. As the light from a distant galaxy approaches us, the universe continues to expand. As the universe expands, the wavelength of that beam of light is stretched. *The farther the light travels, the more it is stretched.* Light from a nearby galaxy arrives at Earth in a relatively short time. During the brief trip, the universe has not expanded by very much, and thus the light waves are not stretched by very much. We observe a small redshift. But light from a remote galaxy spends an enormous time traveling toward Earth. During that time, the universe has had the opportunity to expand significantly. The light waves are substantially stretched out during the course of this lengthy journey and we observe a large redshift. Such redshifts, caused by the expansion of the universe, are called *cosmological redshifts.*

Although we have talked loosely about "nearby" and "remote" galaxies, the effects of the expansion of the universe are detectable only over fairly large distances. Neighborhood galaxies (such as the Andromeda galaxy) are so close to us that the universe does not have the chance to expand significantly during the time that their light travels toward us (only 2 million years for light from the Andromeda galaxy). Indeed, ordinary Doppler shifts caused by the individual motion of a particular nearby galaxy may far outweigh the cosmological redshift caused by the expansion of the universe. For example, the Andromeda galaxy just happens to be approaching us at 260 kilometers per second. This is far greater than the recessional speed caused by the expansion of the universe. Indeed, the spectrum of the Andromeda galaxy exhibits a blueshift. In order to see the cosmological redshift, we must look far beyond our own local neighborhood.

As we look out into space with our most powerful telescopes, we notice that galaxies are not randomly and uniformly scattered across the cosmos. Instead, galaxies occur in *clusters.* There are "rich" clusters that contain as many as a thousand galaxies. The famous Coma cluster (so named because of its location in the constellation of Coma Berenices) shown in Figure 5-1 is a good example of a rich cluster.

And there are "poor" clusters that contain only a handful of galaxies. Our own Galaxy is a member of a poor cluster affectionately called the *Local Group.* Our Local Group contains about twenty galaxies, including M31 (the Andromeda galaxy), M33 (the Triangulum galaxy), and the Small Magellanic Cloud and the Large Magellanic Cloud. There are also several small elliptical galaxies, such as M32 (the companion of M31, shown in Figure 2-4) and NGC 147 (shown in Figure 2-12), along with a few irregulars. The remaining galaxies—nearly a dozen—are all dwarf ellipticals. As mentioned in Chapter 2, these dwarf ellipticals are very difficult to find because they contain so few stars. An alien astronomer in some distant part of the universe probably would not see any of the dwarf ellipticals in the Local Group at all. This suggests that there may be many galaxies (mostly dwarf ellipticals) in distant clusters that escape our scrutiny.

Following the suggestion of George O. Abell at UCLA, astronomers usually prefer to classify clusters of galaxies into one of two categories: *regular* and *irregular*. This distinction is quite similar to the classification of star clusters, as discussed in the first chapter. Recall that there are two types of star clusters: regularly shaped globular clusters (see Figure 1-3) and irregularly shaped open clusters (see Figure 1-2). Similarly, a regular cluster of galaxies (often called a "globular cluster of galaxies") has a distinct spherical symmetry. These clusters are usually very rich, containing at least a thousand members. In addition to the Coma cluster (see Figure 5-1), the Corona Borealis cluster shown in Figure 5-9 is a good example.

An irregular cluster (often called an "open cluster of galaxies") lacks any noticeable spherical symmetry. The Hercules cluster shown in Figure 5-10 is a good example. Whereas regular clusters contain mostly elliptical galaxies, irregular clusters consist of a more democratic mixture of all types of galaxies.

Irregular clusters can be either poor or rich. The Local Group is a good example of a poor irregular cluster. In sharp contrast, the famous Virgo cluster is a fine example of a rich irregular cluster. This huge cluster is so big and so nearby (only 60 million light years away) that it covers roughly 120 square degrees of the sky, mostly in the constellation of Virgo. Only a small portion of the Virgo cluster is shown in Figure 5-11. This view includes the elliptical galaxies M84 and M86. The giant elliptical galaxy M87 mentioned in Chapter 2 (see Figure 2-11) is also a member of this vast cluster. Indeed, 16 of the 39 galaxies listed in Messier's catalog are members of the Virgo cluster.

Relatively speaking, the Virgo cluster is quite nearby. It is the nearest rich cluster. There are also several poor clusters not far from the Local Group. Like the Local Group, these impoverished clusters are usually centered about a large spiral galaxy. Typical examples include the cluster centered about M51 (see Figure 1-10) in Canes Venatici, as well as the cluster centered on M81 (see Figure 3-11) in Ursa Major.

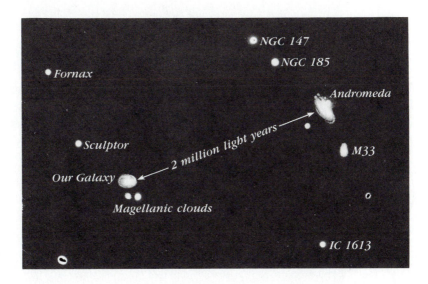

Figure 5-8 The Local Group
*This drawing is a fairly realistic representation of what a remote
alien astronomer looking toward our Galaxy would see. All the
dwarf ellipticals would probably escape detection.*

Figure 5-9   The Corona Borealis Cluster
*This rich cluster in the constellation of Corona Borealis is a good
example of a regular cluster. Regular clusters (often called
"globular clusters") have a distinct spherical symmetry. (Hale
Observatories.)*

Plate 1   The Spiral Galaxy M31 (also called NGC 224) in Andromeda
Copyright by the California Institute of Technology and the Carnegie Institution of Washington. Reproduced by permission from the Hale Observatories.

Plate 2   The Central Regions of the Spiral Galaxy M31
Copyright by the Association of Universities for Research in Astronomy, Inc. Kitt Peak National Observatory.

Plate 3   The Large Magellanic Cloud in Dorado
Copyright by the Association of Universities for Research in Astronomy, Inc. Cerro Tololo Inter-American Observatory.

Plate 4   The Spiral Galaxy M33 (also called NGC 598) in Triangulum
Copyright by the California Institute of Technology and the Carnegie Institution of Washington. Reproduced by permission from the Hale Observatories.

Plate 5   The Spiral Galaxy NGC 253 in Sculptor
Copyright by the California Institute of Technology and the Carnegie Institution of Washington. Reproduced by permission from the Hale Observatories.

Plate 6   The Spiral Galaxy NGC 7331 in Pegasus
Copyright by the California Institute of Technology and the Carnegie Institution of Washington. Reproduced by permission from the Hale Observatories.

Plate 7 The Irregular Galaxy M82 (also called NGC 3034) in Ursa Major
Copyright by the California Institute of Technology and the Carnegie Institution of Washington. Reproduced by permission from the Hale Observatories.

Plate 8 The Peculiar Galaxy NGC 5128 in Centaurus
Copyright by the Association of Universities for Research in Astronomy, Inc. Cerro Tololo Inter-American Observatory.

Figure 5-10 The Hercules Cluster
This cluster in the constellation of Hercules is a good example of an irregular cluster. Irregular clusters (often called "open clusters") have an irregular shape with no apparent spherical symmetry. (Hale Observatories.)

Figure 5-11   A Portion of the Virgo Cluster
This Virgo cluster is a nearby, rich, irregular cluster that covers an area of the sky measuring 10° × 12°. Only a small portion of the entire cluster is shown in this view, which includes the two prominent elliptical galaxies M84 and M86. (Kitt Peak National Observatory.)

From studying the locations of these nearby clusters (including Virgo and the Local Group), Gerard de Vaucouleurs has demonstrated that all together they constitute a huge assemblage called the *Local Supercluster.* Indeed there is a clear tendency for clusters of galaxies to be found in huge swarms called *superclusters.* Typical superclusters, like our own, tend to be about 200 to 300 million light years in diameter. As far as we can tell, superclusters are randomly distributed across the cosmos. In other words, there are no super-duper clusters.

To see the homogeneity of the universe, we must examine the cosmos on scales larger than superclusters. We must gaze across distances greater than hundreds of millions of light years. This is also the scale across which we must look to detect the cosmological redshifts. Galaxies closer than a few hundred million light years are so nearby that their light waves are not appreciably stretched during the brief light-travel time (only a few, scant hundred million years) to us. For example, in Figure 5-5, the galaxy in Virgo is the only galaxy in this illustration closer than a billion light years. The galaxy in Ursa Major, almost exactly a billion light years away, is the first galaxy in this illustration that has a clearly noticeable redshift. For comparison, the galaxy in Hydra, the most distant galaxy shown in Figure 5-5, is 4 billion light years away. Only by peering across these vast distances can we gain profound insight about the creation and fate of the universe as a whole.

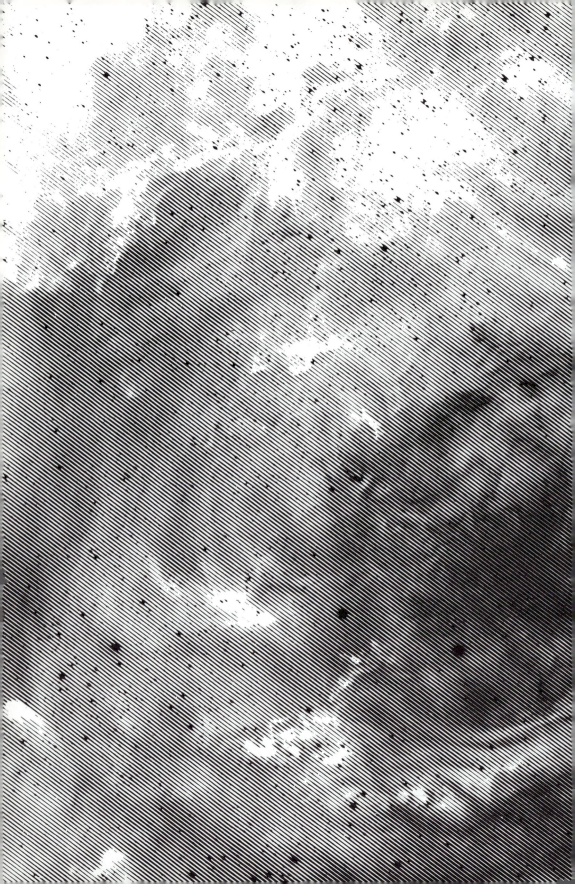

6

The Creation
and the Cosmic
Background

Soon after he had formulated the General Theory of Relativity in 1916, Albert Einstein attempted to use his new theory to construct a theoretical model of the universe. The General Theory of Relativity is a theory about gravity. It tells us how gravity works. According to the old-fashioned ideas of Isaac Newton, gravity is simply a force. In contrast, the relativistic approach explains that gravity curves the fabric of space and time. In the absence of gravity, space and time are flat. In the presence of a gravitational field, space and time become warped. The stronger the gravity, the greater the warping of space-time.

Gravity is the only force in nature that is important on the astronomical scale. Gravity holds the solar system together. Gravity dominates the interaction between stars and galaxies. And gravity dictates the past and future of the universe as a whole. Any mathematical model of the universe must begin with a theory of gravity. Observations have proved that the General Theory of Relativity is a more accurate description of gravity than the classical, Newtonian approach.

According to Newton's ideas, the universe must be static and infinite. If the universe were finite, the mutual gravitational forces between all the stars and galaxies would cause the universe to collapse. Only if the universe is infinite and unchanging can a classical cosmos survive. As we saw at the beginning of the previous chapter, these ideas lead directly to Oblers paradox.

Although he had a more accurate theory of gravity, Einstein was frustrated in his attempts to construct mathematical models of the universe. No one had any suspicion of large-scale motions in the universe, so Einstein labored to formulate static models. His efforts met with failure; his mathematical universes would always collapse. Instead of having faith in his calculations (his formulas were really telling him that we do *not* live in a static universe), Einstein began to doubt the validity of his equations. Accordingly, he added to his equations a special number, called the *cosmological constant,* that would prevent his static universe from collapsing. This cosmological constant represents a long-range cosmic force (not detectable over short distances in a laboratory) that literally holds the universe up.

By forcing his models of the universe to be static, Einstein missed the opportunity to anticipate Hubble's discovery by at least a decade. By 1929, it was clear that we really live in an expanding universe. No wonder static models never seemed to work out. In later years, Einstein lamented that including the cosmological constant in the basic field equations of general relativity was "the biggest mistake of my life."

As we saw in the previous chapter, the Hubble law is a correlation between redshift and distance. As we look toward increasingly distant galaxies, we see correspondingly larger redshifts. This relationship is most easily displayed in a graph such as Figure 5-6. Although clever physicists have occasionally proposed alternative ideas,* the most straightforward interpretation of the Hubble law is that we live in an expanding universe. As radiation from a distant galaxy approaches us, the universe is expanding, and consequently the light waves become stretched.

This interpretation of the Hubble law tells us that the size of the redshift of a particular galaxy equals the amount by which the universe has expanded during the time that the light has been traveling toward us. For example, the galaxy called 3C295 shown in Figure 6-1 has a redshift of $z = 0.46$. The wavelengths of all the light we receive from 3C295 are 46 percent longer than usual. This means that the universe is now 46 percent larger than it was when the light left 3C295. More precisely, the distances between widely separated clusters of galaxies have increased by 46 percent during the time that light was traveling from 3C295 to us.

Because clusters of galaxies are now getting farther and farther apart, there must have been a time when they were very close together. In fact, there must have been a time in the distant past when all of the matter and energy in the universe was crowded together in a very dense state. From knowing how fast the universe is expanding, we can extrapolate backward in time to the moment when the universe was infinitely dense. That moment occurred roughly 20 billion years ago. It was the moment of creation.

Because we live in an expanding universe, and because this expansion began about 20 billion years ago, a vast explosion

---

*For a discussion of some of these clever, nonstandard cosmologies, you are referred to *Relativity and Cosmology*, 2nd ed. (Harper & Row, New York, 1977) by the author of this book.

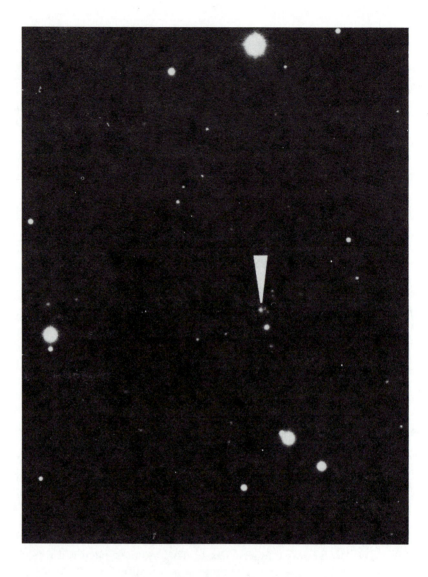

Figure 6-1   A Very Distant Galaxy
This galaxy, called 3C295, is the brightest member of a very remote
cluster of galaxies. The light from 3C295 exhibits a redshift of 46
percent. This redshift corresponds to a velocity of 36 percent of the
speed of light. (Hale Observatories.)

throughout all space must have occurred to start the expansion. This cosmic explosion is called the *Big Bang.* The redshifts of the galaxies attest to this universal detonation from which the cosmos was born.

It is important to try to understand exactly what is meant by the Big Bang. It was *not* like the explosion of a bomb that scatters fragments outward into otherwise empty space. Twenty billion years ago, the galaxies were *not* clumped together in a large lump somewhere in the universe.

At the moment of creation, the entire universe—*all space*—was filled with matter and energy at infinite density. The universe does not have a center. The universe does not have an edge. Therefore it could *never* have had a center or an edge. The entire, limitless universe participated in the Big Bang. The explosion occurred throughout *all* space at the beginning of time. Indeed, *the Big Bang was an explosion of space itself.* From that moment on, the amount of space between widely separated locations began increasing. Even today, the amount of space between widely separated clusters of galaxies continues to increase. This is why we observe redshifts. As light from a distant galaxy approaches us, the amount of space between us and that galaxy increases, and thus the wavelengths become stretched.

This interpretation of the Hubble law follows directly from the General Theory of Relativity. It is a more rigorous, more precise explanation of the redshifts of galaxies than the raisin cake analogy discussed in the previous chapter. The raisin cake analogy, which is based on old-fashioned classical ideas, helped indicate what is going on. It helped point the way to the correct interpretation. But for the correct description we must use our best theory of gravity—the General Theory of Relativity—which tells us that the amount of space between widely separated galaxies can change over the years. The relativistic approach tells us what it truly means to live in an expanding universe.

In the late 1940s, American physicists George Gamow, Ralph Alpher, and Robert Herman began thinking about conditions in the early universe. They argued that it must have been much hotter billions of years ago when the universe was much denser and much more compressed than it is today. Under these hot, dense conditions, thermonuclear reactions could have occurred throughout the universe—the same reactions that presently occur at the centers of stars like the sun. At the sun's center, the nuclei of hydrogen atoms are fused together to produce helium. Gamow, Alpher, and Herman argued that the "burning" of hydrogen shortly after the Big Bang could account for the primordial helium that we find in the universe. An important and prophetic conclusion of their work is that the universe must have been very hot billions of years ago and that some of this heat should still be around today.

In the early 1960s, a young theoretical physicist at Princeton University, P. J. E. Peebles, began reexamining ideas about the early universe in greater detail. Peebles's calculations soon revealed the importance of a high temperature shortly after the Big Bang. Without the intense radiation associated with high temperatures, thermonuclear reactions would have proceeded at a furious rate. All the hydrogen would have rapidly been converted into helium, which soon would have been fused into heavier elements. Obviously this did not happen. The primordial gas left over from the Big Bang consists of about 75 percent hydrogen and 25 percent helium, by weight. We know that these are the only gases out of which the oldest stars are made. No heavier elements survived the birth of the universe. To ensure this, Peebles argued, intense radiation following the Big Bang fragmented heavier nuclei as soon as they formed. The entire early universe must have been filled with a bath of intense radiation at a temperature of several hundred million degrees. This was the *primordial fireball.*

When physicists speak of light or radiation, they are usually talking about *photons.* A photon may be thought of as a very tiny piece of radiation, just as an atom is a very tiny piece of matter. Ever since the early 1900s, scientists have realized that radiation actually consists of photons—often so tiny and so innumerable that they cannot be detected individually.

The energy carried by a photon depends only on the photon's wavelength. The shorter the wavelength, the higher the energy. Of all types of radiation, gamma rays have the shortest wavelength. A gamma ray photon therefore carries a sizable amount of energy. Indeed, gamma rays from a nuclear reactor can pass through several feet of lead or concrete. X-ray photons have slightly longer wavelength than gamma rays. Hence, X-ray photons carry slightly less energy than gamma rays. X rays pass easily through your body but not through lead. Of all radiations, radio waves have the longest wavelength. Of all types of radiation, a radio wave photon carries the least amount of energy.

An object at a particular temperature emits photons at a predominant wavelength. A cool object at a few degrees above absolute zero emits long-wavelength, low-energy radio wave photons. A warmer object several hundred degrees above absolute zero (such as hot coals in a barbecue) emit moderate-energy photons at intermediate wavelengths. And very hot objects at several million degrees are copious sources of high-energy, short-wavelength photons, such as X rays and gamma rays. According to Peebles, the high-energy gamma ray photons that pervaded the hot, early universe fragmented any heavy nuclei as soon as they formed. Without this primordial fireball of energetic gamma ray photons, all of the original hydrogen would have been converted into heavy elements, such as iron.

During the earliest moments of the universe, when densities were enormous and temperatures were high, the universe must have been opaque. Gamma ray photons could travel only a tiny distance before hitting a particle or another photon.

But as the universe expanded, the wavelengths of all the photons in the primordial fireball became longer and longer. As the wavelengths increased, the energy carried by each photon decreased. As the energy carried by the photons decreased, the temperature of the primordial fireball declined. The universe began to cool off. The declining temperature of the primordial fireball with age is displayed in Figure 6-2.

Hydrogen is by far the most abundant substance in the universe. A hydrogen atom — the simplest and lightest atom in nature — consists of a negatively charged electron orbiting a positively charged proton, like a tiny atomic solar system. During the first 700,000 years of the universe, it was too hot for electrons and protons to combine to form hydrogen atoms. Constant bombardment by

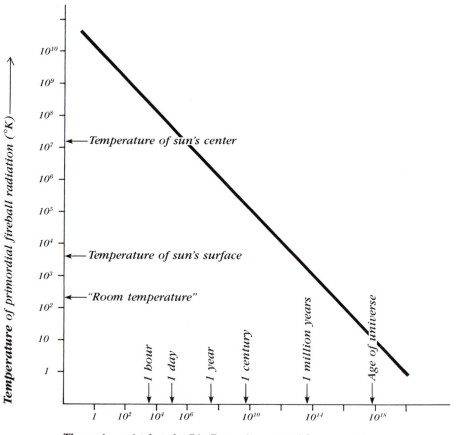

Figure 6-2 The Cooling of the Primordial Fireball
As the universe expands, the wavelengths of all the photons in the primordial fireball become stretched. As the wavelengths increase, the energy carried by each photon decreases. Consequently, the temperature of the fireball decreases.

high-energy photons broke apart the hydrogen atoms as soon as they formed.

Free electrons floating around in the hot, early universe played an important role in keeping the universe opaque. Free electrons easily and efficiently scatter photons. A photon could not travel very far without getting knocked around by all the loose electrons.

When the universe was about 700,000 years old, the primordial fireball had cooled off to the extent that hydrogen atoms could form. No longer was there an ample supply of short-wavelength, high-energy photons to keep knocking the protons and electrons apart. The expansion of the universe had stretched the wavelengths of photons, thereby substantially reducing their potency. Protons and electrons were finally at liberty to cling together as hydrogen atoms.

The formation of hydrogen gas across the cosmos was a crucial event in the history of the universe. For the first time, there was not an ample supply of free electrons to scatter and deflect the photons of the fireball. For the first time, radiation of the primordial fireball could stream across space virtually unimpeded. The universe had become transparent!

The universe became transparent 700,000 years after the Big Bang. At that moment, the temperature of the primordial fireball was about 3,000°K. Since that time, the universe has expanded by a factor of roughly 1,000. Consequently, the wavelengths of photons left over from the primordial fireball have all been stretched by a factor of 1,000. They have all suffered a redshift of $z = 1,000$. This means that the temperature of the radiation from the primordial fireball has decreased by a factor of 1,000. It should now be about 3 degrees above absolute zero.

In 1964, while Peebles was just beginning to formulate his ideas about the primordial fireball, two radio astronomers were attempting to detect microwaves from outer space. Arno A. Penzias and Robert W. Wilson were making these observations with a new and unusual antenna that had been built at Bell Telephone Laboratories, in Holmdel, New Jersey, for satellite communications. This "horn antenna" is shown in Figure 6-3. Much to their surprise, Wilson and Penzias found that their antenna picked up static all

Figure 6-3   The Holmdel Horn Antenna
*Using this horn antenna, Robert Wilson and Arno Penzias detected
the cosmic microwave background in 1965. For this important
discovery they were awarded the Nobel Prize in Physics in 1978.
(Courtesy of Bell Laboratories.)*

across the sky. Regardless of the direction in which they pointed their antenna, they detected microwaves—photons of moderately long wavelength corresponding to a theoretical temperature at about 3 degrees above absolute zero. Wilson and Penzias were at first puzzled by this pervasive background of microwaves, while rumors of the discovery spread to Princeton. It was soon apparent that Wilson and Penzias had detected the highly redshifted radiation left over from the primordial fireball. Figuratively speaking, humanity had heard the echo of the Big Bang.

Wilson and Penzias had made only one observation at one particular wavelength (7.35 centimeters). Over the next few years, teams of radio astronomers labored to measure the intensity of the microwave background at various wavelengths. This was important because if the microwave background is the cooled-off radiation from the primordial fireball, then all the measurements over a range of wavelengths should conform to a particular theoretical curve for a particular temperature. By the early 1970s, enough data had been collected at enough wavelengths to satisfy everyone; the current temperature of the fossil radiation left over from the primordial fireball is 2.76 degrees above absolute zero. Figure 6-4 shows the data from measurements at various wavelengths, along with the theoretical curve that best fits these data.

For more than a decade after the initial discovery, everyone believed that the cosmic microwave background was completely uniform across the sky. No one had been able to detect the slightest variation from one part of the sky to another. But beginning in December 1976, a team of physicists headed by Richard A. Muller flew sensitive microwave detectors on board a U-2 airplane operated by NASA. Ten flights were made during the next twelve months at altitudes in excess of 50,000 feet. As they scanned large sections of the sky with unprecedented accuracy, Dr. Muller and his colleagues found a small but regular variation in the temperature of the microwave background. The temperature of the microwave background is a little hotter (by about one three-hundredths of a degree) in the direction of the constellation of Leo. In exactly the opposite direction, toward the constellation of Aquarius, the microwave back-

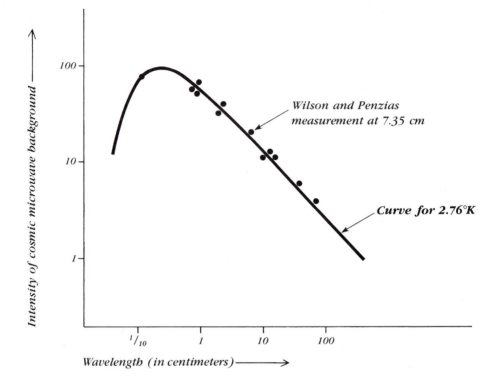

Figure 6-4   The Cosmic Microwave Background
All space is filled with a background of microwaves. By measuring the intensity of these microwaves at many wavelengths, scientists have concluded that the temperature of this cooled-off radiation from the primordial fireball is now 2.76 degrees above absolute zero.

ground is a little colder, by exactly the same amount. Toward Leo we see a slightly warmer background, as evidenced by slightly shorter-than-expected wavelengths. Toward Aquarius we find a slightly cooler background, as evidenced by slightly longer-than-expected wavelengths. In between these two extremes, the temperature of the background declines in a smooth and regular fashion from the high in Leo to the low in Aquarius.

The microwave background is a universal bath of radiation that permeates all space. If we happen to be moving through space relative to this cosmic background, then we should observe a Doppler shift in the wavelengths of microwaves across the sky. Shorter-than-usual wavelengths should be detected in the direction toward which we are going. And longer-than-usual wavelengths should be found in the direction from which we came, as diagrammed in Figure 6-5.

This is the obvious interpretation of the U-2 data. We have discovered the *absolute motion* of the earth through the universe. Our planet is rushing toward a point in the constellation of Leo at a speed of 400 kilometers per second.

But, of course, the earth goes around the sun and the sun revolves about the center of the Galaxy. By sorting out these well-known motions, we can calculate the absolute motion of the entire Galaxy across the cosmos. Our Galaxy is rushing across space at the surprisingly high speed of 600 kilometers per second (roughly 1⅓ million miles per hour).

We know that our Galaxy's velocity with respect to other galaxies in the Local Group is quite small. For example, the relative speed between our Galaxy and the Andromeda galaxy is only 80 kilometers per second. Consequently, the entire Local Group must share our Galaxy's high speed. The entire Local Group must be rushing across space at roughly a million miles per hour.

As mentioned in the previous chapter, the nearest rich cluster is in Virgo. The Virgo cluster, the Local Group, and several other nearby clusters comprise the Local Supercluster. The relative speed between our Local Group and the other clusters in the Local Supercluster is not large. Consequently, because our absolute veloc-

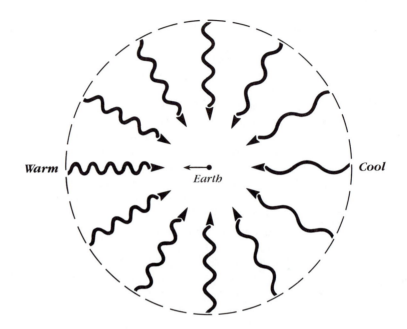

Warm   *Earth*   **Cool**

Figure 6-5   Doppler Shift of the Microwave Background
The microwave background is a uniform bath of radiation that permeates the entire universe. If the earth is moving through this cosmic radiation field, then we should detect Doppler shifts across the sky. We should see shorter wavelengths (and thus a higher temperature) in the direction toward which we are moving. And longer wavelengths (corresponding to a lower temperature) should be observed in the direction from whence we came. (Adapted from R. A. Muller.)

ity is high, the entire Local Supercluster must be rushing across space.

This conclusion is troublesome. If the entire Local Supercluster is in motion with respect to the cosmic microwave background, then we must have traveled a considerable distance over the past 15 to 20 billion years. In principle, we should have caught up to that portion of the universe whose speed (just because of the natural expansion of the universe) equals the speed of the Local Supercluster. But apparently we are still moving at a measurable velocity.

Perhaps a better interpretation of the data is that the Local Supercluster is rotating. Perhaps our Local Group as a whole is simply orbiting the common center of mass of the entire Supercluster, just as our planet orbits the sun. Perhaps the tiny temperature variations of the microwave background across the sky are our first hint of organized orbital motions on an inconceivably colossal scale.

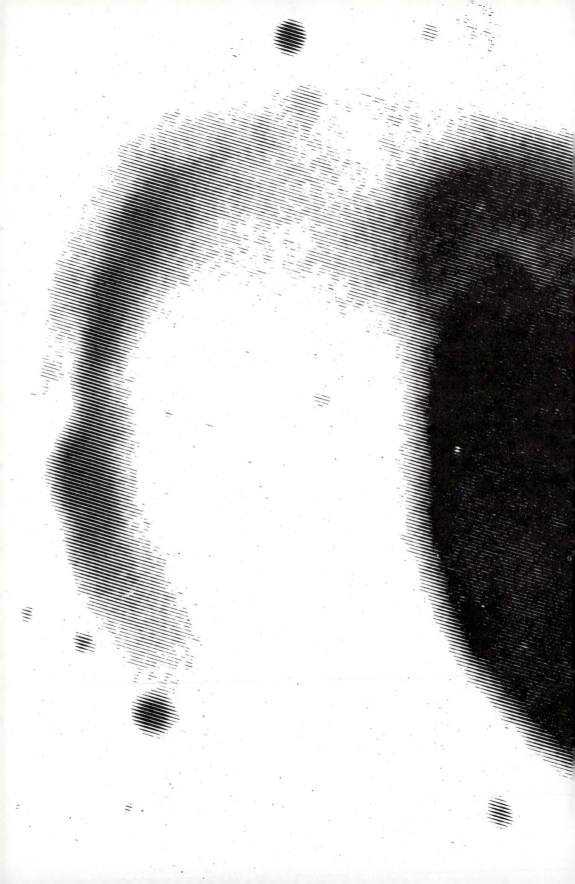

7

Quasars

The birth of radio astronomy in the late 1940s surely ranks among the most important scientific developments of this century. Prior to that time, everything that humanity knew about the distant universe was gleaned from optical observations alone. But as soon as astronomers began building radio telescopes, we had the opportunity to view the universe in a wavelength range far removed from ordinary visible light. As mentioned in Chapter 3, the vista provided by a radio telescope is significantly different from what we see through ordinary optical telescopes. Many unexpected and surprising discoveries were destined to ensue from our newborn ability to examine the previously invisible universe.

The first radio telescope was built during World War II by an amateur astronomer, Grote Reber, in his backyard in Illinois. By 1944, Reber had detected strong radio emissions from the constellations of Sagittarius, Cassiopeia, and Cygnus. The existence of these sources of radio waves was confirmed in 1946 by a team of astronomers with a brand-new radio telescope in England. Unfortunately, they could not establish precise positions. For example, the location of the source in Cygnus, nicknamed Cygnus A, could not be pinned down with the relatively crude instruments of the day.

A major advance occurred in 1954, when British and Australian astronomers succeeded in building the first radio interferometers. As I mentioned in Chapter 3, radio waves have much longer wavelengths than ordinary visible light. Consequently, in order to get a sharp radio picture, you must use a huge radio telescope. However, it is possible to wire several small radio telescopes together to give the effect of a large telescope. Such an assemblage is called a *radio interferometer.* One of these newfangled interferometers allowed Graham Smith at Cambridge, England, to nail down Cygnus A.

As soon as the location of Cygnus A had been established, Walter Baade and Rudolph Minkowski used the giant 200-inch optical telescope on Palomar Mountain to see if any strange-looking objects might be found at that position. Indeed, they discovered a peculiar galaxy at the location of Cygnus A. A photograph of this galaxy is shown in Figure 7-1.

The peculiar galaxy associated with Cygnus A is very dim. Nevertheless, Baade and Minkowski managed to photograph its spectrum, which showed a number of bright spectral lines among the

Figure 7-1   Cygnus A (also called 3C405)
*This strange-looking galaxy was discovered at the location of
Cygnus A. This galaxy has a substantial redshift (z = 0.057), which
means that it must be far away (about 1 billion light years from
Earth). Because Cygnus A is one of the brightest radio sources in the
sky, the energy output of this remote galaxy must be enormous.
(Hale Observatories.)*

colors of the rainbow. To everyone's surprise, all of these spectral lines were shifted by 5.7 percent toward the red end of the spectrum. A redshift of $z = 0.057$ corresponds to a speed of 17,000 kilometers per second, which, in turn—according to the Hubble law—corresponds to a distance of 1 billion light years!

This result was astounding. Cygnus A is very far away. Yet it is one of the brightest sources of radio waves in the sky. Although Cygnus A is hardly visible through the 200-inch telescope at Palomar, its radio waves can be picked up by amateur astronomers with backyard equipment. The energy output in radio waves of Cygnus A must therefore be enormous. Indeed, Cygnus A shines with a radio luminosity that is 10 million times as bright as ordinary galaxies like M31 in Andromeda. Obviously astronomers had stumbled onto something quite extraordinary.

During the late 1950s and early 1960s, radio astronomers were busy making long lists of all the radio sources they were finding across the sky. One of the most famous lists is called the *Third Cambridge Catalogue* (the first two catalogs produced by the British team were filled with inaccuracies) and was published in 1959. It listed 471 sources. Even today, astronomers often refer to these sources according to their "3C numbers." With the discovery of the extraordinary luminosity of Cygnus A (also called 3C405, because it is the 405th source on the Cambridge list), astronomers were highly motivated to find out if any other sources in the 3C catalog had similar extraordinary properties.

Some of the 3C sources turned out to be associated with nearby objects in our own Galaxy. For example, the radio source that Reber had discovered in the constellation of Taurus (called Taurus A, or 3C144) is the famous Crab Nebula, the first object on Messier's list. Other sources turned out to be galaxies, many of which have very disturbed or peculiar appearances. The giant elliptical galaxy M87 (called 3C284) and the irregular galaxy M82 (called 3C231) are good examples. Several of these fascinating cases will be discussed in the next chapter.

Many of the 3C sources were apparently not associated with anything unusual—perhaps a star, perhaps nothing. One interesting case was 3C48. In 1960, Allan Sandage of Hale Observatories had discovered a "star" at the location of this source. A photograph of

3C48 is shown in Figure 7-2. Everyone realized that ordinary stars simply cannot produce enough radio waves to be detected over galactic distances. So, 3C48 must be peculiar. Indeed, its spectrum showed a series of spectral lines that no one could identify. Although 3C48 was clearly an oddball, it was generally agreed that astronomers had merely stumbled upon another strange star here in our own Galaxy.

The mystery deepened in 1962 with the discovery of another one of these radio "stars." In a continuing effort to identify sources in the *Third Cambridge Catalogue,* astronomers at the Australian National Radio Observatory observed several lunar occultations of 3C273. A lunar occultation occurs any time the moon eclipses a background object. Because of its location in the constellation of Virgo (one of the constellations of the zodiac), 3C273 was scheduled to be covered by the moon three times during the late summer and fall of 1962. Because they knew exactly where the moon was in the sky, the Australian astronomers managed to determine the exact location of 3C273 by simply noting the times when the radio waves were blocked by the moon. Once again, a "star" was found at the position of 3C273.

A superb photograph of 3C273 is shown in Figure 7-3. Notice that there is a strange-looking luminous "jet" to one side of the starlike object. Because a visible object had been identified, astronomers could use a spectrograph to obtain a spectrum. Once again, the spectrum of the "star" contained a series of bright spectral lines among the colors of the rainbow that no one could identify.

The central difficulty with identifying the spectral lines in the spectra of these starlike objects was that everybody believed that 3C48 and 3C273 were nearby peculiar stars here in our own Galaxy. After all, they certainly looked like stars.

In 1963, Maarten Schmidt at Caltech was examining the spectrum of 3C273 when he realized that four of the brightest spectral lines were positioned relative to each other with exactly the same spacing as four very familiar spectral lines of hydrogen. But 3C273's lines were located quite far from the usual positions for hydrogen lines among the colors of the rainbow; they had much longer wavelengths than usual. If 3C273 really is a nearby star, then these lines could not possibly be the familiar lines of hydrogen. So perhaps

Figure 7-2 The Quasar 3C48
For several years astronomers believed erroneously that this object was simply a nearby peculiar star that just happens to emit radio waves. Actually, the redshift of this starlike object is so great (z = 0.367) that, according to the Hubble law, it must be roughly 5 billion light years away. (Hale Observatories.)

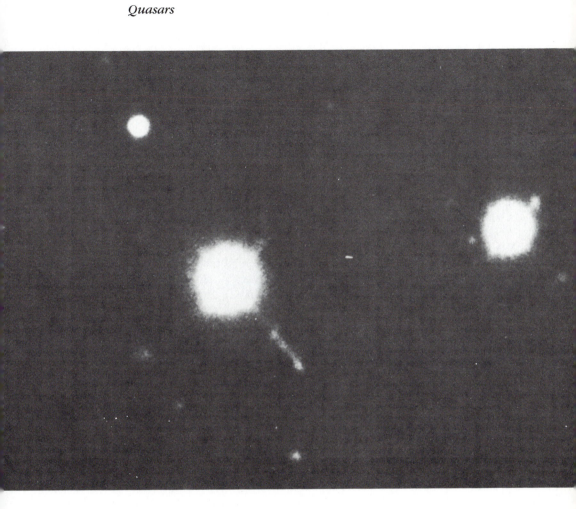

Figure 7-3   The Quasar 3C273
*This highly enlarged view shows the starlike object associated with
the radio source 3C273. Notice the luminous jet to one side of the
"star." By 1963, astronomers discovered that the redshift of this
"star" is so great ($z = 0.158$) that its distance according to the
Hubble law must be nearly 3 billion light years from Earth. (Kitt
Peak National Observatory.)*

3C273 is *not* a nearby star. Schmidt pursued this hunch and promptly identified all four spectral lines as being hydrogen lines that had suffered an enormous redshift. They were all shifted by almost 16 percent from their usual wavelengths. This corresponds to a speed of 45,000 kilometers per second (that is, 15 percent of the speed of light). According to the Hubble law, this huge redshift implies an incredible distance to 3C273: nearly 3 billion light years!

Obviously, 3C273 is not a star. But because of its starlike appearance, it was called a *quasi-stellar object.* This term was soon shortened to *quasar.*

The spectrum of 3C273 is shown in Figure 7-4. Actually, this graph was made from the spectroscopic plate on which the spectrum had been photographed. The four hydrogen lines are identified. Notice that the spectral lines are brighter than the average intensity of the background rainbow of colors. Bright lines are called *emission lines* (as opposed to *absorption lines,* which are darker than the background intensity of the rainbow). In Chapter 5 (see Figure 5-3), I explained that absorption lines are produced when white light passes through cool gas. Atoms in the gas absorb light at specific wavelengths, thereby leaving blank spaces among the colors of the rainbow. But in order to produce bright emission lines, the gas must be hot rather than cool. The atoms in the gas are "excited" and emit radiation at specific wavelengths rather than absorb it. All quasars and most peculiar galaxies (like Cygnus A and many of the strange galaxies we will discuss in the next chapter) exhibit strong emission lines in their spectra. It's a sure sign that something unusual is going on.

The identification of four hydrogen lines in the spectrum of 3C273 permitted Schmidt to recognize the remaining spectral lines as being due to carbon and oxygen. They all had the same huge redshift. Armed with these identifications, two other Caltech astronomers, Jesse Greenstein and T. A. Matthews, reexamined the mysterious spectrum of 3C48. Indeed, the only reason its spectrum had been mysterious is that no one had considered the possibility of huge redshifts. But now everything began falling into place. Greenstein and Matthews promptly identified all the spectral lines of 3C48 as having suffered a redshift of $z = 0.367$. That corresponds to a velocity of nearly one-third the speed of light. According to the Hubble law, that

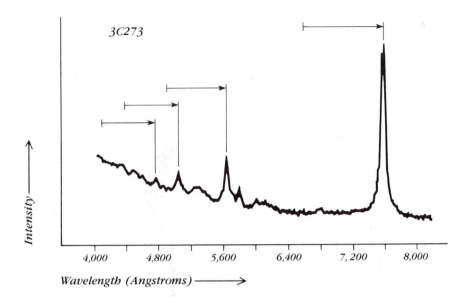

3C273

Intensity →

4,000   4,800   5,600   6,400   7, 200   8,000

Wavelength (Angstroms) →

Figure 7-4   The Spectrum of 3C273
The spectrum of 3C273 exhibits four bright emission lines due to hydrogen. A graph like this, called a "tracing," can be made from the tiny spectroscopic plate on which the spectrum has been photographed. The arrows indicate how far the spectral lines are redshifted.

places 3C48 nearly *twice* as far away as 3C273—over 5 billion light years away!

The revelation that quasars possess huge redshifts was a major breakthrough. The floodgates were about to be opened. During 1964, eight more quasars were identified. One of them, 3C147 shown in Figure 7-5, has a redshift of $z = 0.545$. This corresponds to a speed of 123,000 kilometers per second (41 percent of the speed of light). According to the Hubble law, the distance to 3C147 must be 7 billion light years.

Figure 7-5   The Quasar 3C147
This quasar was identified in 1964. Its spectral lines are redshifted by 54 percent from their usual wavelengths. This redshift corresponds to a speed of 41 percent of the speed of light. According to the Hubble law, 3C147 is therefore at a distance of 7 billion light years. (Hale Observatories.)

This record did not stand for long. In 1965, Maarten Schmidt measured a redshift of $z = 2.012$ for the quasar 3C9 shown in Figure 7-6. No galaxy had ever been seen with a redshift this big. The spectral lines that Schmidt identified in the case of 3C9 had been redshifted all the way from normally invisible ultraviolet wavelengths into the visible part of the spectrum. According to the Special Theory of Relativity (refer to Figure 5-4), this redshift implies a speed of 80 percent of the speed of light. According to the Hubble law, an object with a redshift this big must be at a distance of 14 billion light years.

Hundreds of quasars have been discovered since those pioneering days back in the early 1960s. Enormous redshifts are commonplace. There are even a few quasars known to have redshifts greater than 3. For example, the quasar OH471 shown in Figure 7-7 has a redshift of $z = 3.4$. That corresponds to a speed of slightly over 90 percent of the speed of light, which in turn (according to the Hubble law) implies a distance of 16 billion light years. Because the universe is roughly 20 billion years old, OH471 must have been shining brilliantly only 4 billion years after the Big Bang. Its light has taken 16 billion years to reach us. When we look at OH471, we are seeing one of the most ancient objects in the universe.

As one quasar after the next was being discovered across the sky, astronomers began realizing that they had uncovered a perplexing dilemma. Traditionally, the biggest and brightest objects in the universe are galaxies. A typical large galaxy like our own or M31 contains a hundred billion stars and shines with a luminosity of 10 billion suns. The most gigantic and most luminous galaxies (such as the supergiant elliptical M87) are only ten times as bright, shining with the brilliance of a hundred billion suns. But no galaxy has ever been seen with a redshift greater than about $z = 0.6$. A redshift of 60 percent corresponds to a distance of roughly 7 billion light years. Beyond that distance, galaxies—even the brightest—are just too faint to be detected. Most ordinary galaxies are too dim to be seen at half that distance. Although it is difficult to find high-redshift galaxies, we have no trouble stumbling over numerous quasars, most of which have huge redshifts. If quasars are at the vast distances indicated by their redshifts, then quasars must be incredibly luminous. They must be far more luminous than galaxies. Indeed, a typical quasar must

Figure 7-6   The Quasar 3C9
*This faint quasar was discovered in 1965 and has a redshift of
z = 2.012. This corresponds to a speed of 80 percent of the speed
of light. According to the Hubble law, 3C9 is at a distance of
14 billion light years. (Hale Observatories.)*

142

Figure 7-7   The Quasar OH471

This quasar has one of the biggest redshifts ever discovered ($z = 3.4$). This redshift corresponds to a speed slightly over 90 percent of the speed of light. According to the Hubble law, OH471 must be 16 billion light years away. (Copyright by the National Geographic Society—Palomar Sky Survey. Reproduced by permission from the Hale Observatories.)

shine with a brilliance 100 times as bright as the brightest galaxies we have ever seen.

It seems as though quasars are by far the most luminous objects in the universe. But that is only part of the dilemma. In the mid-1960s, several astronomers began rummaging through stacks of old photographs to see if some of the newly identified quasars had inadvertently been photographed in the past. Indeed they had. For example, 3C273 was found on numerous photographs, including one taken in 1887. By carefully examining the images of quasars on these old photographs, astronomers found that quasars fluctuate in brightness. Occasionally they flare up. For example, old photographs of 3C279 revealed that this quasar flared up twice, once in 1936 and again in 1937. During each outburst, the brightness of 3C279 increased by a factor of 25. Because of the enormous distance to 3C279, this quasar must have been shining with a brilliance 10,000 times as great as that of the Andromeda galaxy at the peak of each of these outbursts.

Of course, that is an astounding energy output. But more important, fluctuations in brightness allow us to place strict limits on the maximum sizes of quasars. Specifically, an object cannot vary in brightness faster than the light-travel time across that object. For example, an object that is 1 light year in diameter cannot vary in brightness faster than once a year.

Many quasars vary their brightness over periods of a few months to a year. Some quasars fluctuate from night to night. This rapid flickering means that quasars are small. The energy-emitting region of a quasar — the "powerhouse" that blazes with the luminosity of a hundred galaxies — cannot be more than a few light days across. This reveals the full scope of the dilemma surrounding quasars. If quasars are at the huge distances indicated by their redshifts, then they must be producing the luminosity of a hundred galaxies in a volume not much bigger than our solar system!

The magnitude of this enigma — that something so small could produce so much energy — has prompted some astronomers to consider the possibility that quasars are *not* located at the vast distances inferred from their redshifts. If quasars do not obey the Hubble law, then they could be much nearer to us than their redshifts would have us believe. And if quasars are nearby, then the "energy problem" goes away. But a new problem arises. If quasars are nearby, then what causes their big redshifts? Standard processes in physics cannot explain how a nearby quasar can have a high redshift. From the traditional viewpoint, the only thing that makes sense is the Hubble-law interpretation: a big redshift means a big distance. Consequently, maverick astronomers who profess the anti-Hubble-law interpretation of quasar redshifts often argue that we are at the brink of discovering *new* laws of physics. This "new physics" would then explain how quasars can be nearby and have large redshifts.

This nontraditional approach to the quasar dilemma is not popular. Most astronomers feel that the Hubble law is a valid indicator of distance. The traditional astronomer believes that quasars are indeed located at the enormous distances indicated by their huge redshifts. These astronomers accept the challenge of trying to explain how the luminosity of a hundred galaxies can be generated in a volume not much bigger than our solar system. (More about this in the next chapter!)

But the maverick astronomers have produced some interesting observations that might indicate something wrong with the traditional approach. The real issue is, Where are the quasars? Unfortunately, no one has been able to devise an independent and indisputable method of measuring the distances to quasars. For example, nobody has been able to discover the equivalent of Hubble's cepheid variables that settled the Shapley-Curtis debate back in the 1920s. Nevertheless, Halton C. Arp of Hale Observatories has found a number of cases where the Hubble law is apparently not obeyed. For example, in 1972, Dr. Arp found a quasar that seems to be attached to a nearby galaxy by a luminous bridge of gas. This case is shown in Figure 7-8. The galaxy (NGC 4319) has a low redshift ($z = 0.006$), corresponding to a speed of only 1,800 kilometers per second. According to the Hubble law, NGC 4319 is only 100 million light years away. But the quasar (Markarian 205) has a much higher redshift ($z = 0.07$), corresponding to a speed of 21,000 kilometers per second. (Incidentally, this quasar derives its name from being the 205th object in a catalog compiled by the Russian astronomer B. E. Markarian in the early 1970s.) If Markarian 205 is really connected to NGC 4319, then the quasar would have to be much closer to us than the distance implied by

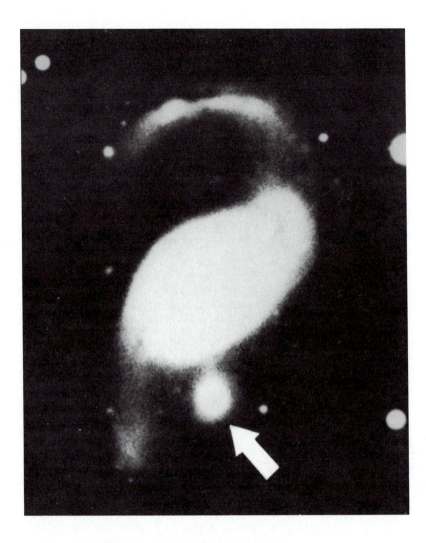

Figure 7-8   NGC 4319 and Markarian 205
The galaxy NGC 4319 has a small redshift corresponding to the relatively nearby distance of only 100 million light years. The quasar (indicated by the arrow) has a redshift eleven times as large as the galaxy. But the quasar appears to be attached to the galaxy. If these two objects are really connected, this case would be a major violation of the Hubble law. (Courtesy of H. C. Arp; Hale Observatories.)

its redshift and the Hubble law. Indeed, this would constitute a blatant violation of the Hubble law.

Traditional astronomers argue that this entire situation is a case of "chance alignment." The galaxy NGC 4319 is where it belongs: at the distance indicated by its redshift. The quasar Markarian 205, which has a redshift eleven times as large as that of the galaxy, is where it belongs: eleven times as far away. They only *look* connected because they just happen to lie nearly in the same direction in the sky.

Arp and his cohorts counter these objections by pointing to many cases where galaxies with widely differing redshifts appear to be located together in space. Three fine examples of these *discrepant redshifts* are shown in Figures 7-9, 7-10, and 7-11.

In 1959, the Soviet astronomer B. A. Vorontsov-Velyaminov published a catalog listing many strange clusters and groupings of galaxies. The 172nd object in that catalog is shown in Figure 7-9. The fact that all five galaxies in the chain are evenly spaced in a straight line suggests that they are all located together. But one of the galaxies has a much higher redshift than the other four. According to the traditional viewpoint, the high redshift galaxy must be much farther away than the other four galaxies. It must be a background galaxy that just happens to appear in a gap in the nearby chain of four galaxies. But Arp finds it impossible to believe that a remote, background galaxy just happened — *totally by chance* — to be positioned with the alignment and spacing we see in Figure 7-9.

Figure 7-10 shows the famous Seyfert's Sextet, a grouping of six galaxies discovered by the American astronomer Carl Seyfert in 1954. They all have roughly the same brightness and appear to be clustered together. But one galaxy has a redshift that is more than four times as large as the redshifts of the others. Once again, the traditional viewpoint argues that the high-redshift galaxy is a background object, four times as far away as the other galaxies. But the apparent tight grouping that we see in Figure 7-10 makes us wonder if something might be wrong with the traditional approach.

In 1877, the French astronomer M. E. Stephan discovered the small cluster of galaxies shown in Figure 7-11. Four of the galaxies in Stephan's Quintet have nearly the same redshift. But in 1961, the husband-and-wife team of Geoffrey and Margaret Burbidge discov-

Figure 7-9   The Chain of Galaxies VV172
Four of the galaxies in this cluster have nearly the same redshift
(z = 0.05). But the remaining galaxy (indicated by an arrow) has
a much higher redshift (z = 0.12). (Courtesy of H. C. Arp; Hale
Observatories.)

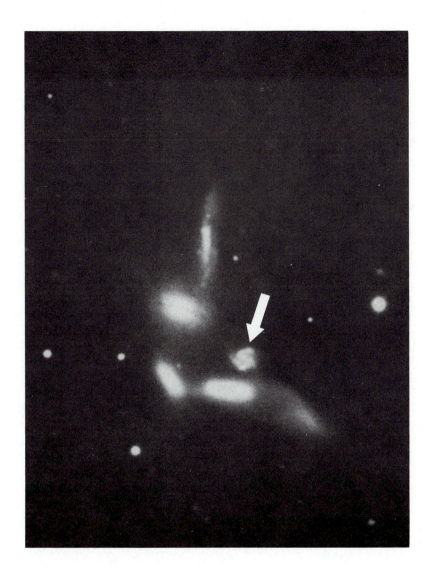

Figure 7-10 Seyfert's Sextet
Five of the galaxies in this cluster have nearly the same redshift
(z = 0.015). But the remaining galaxy (indicated by an arrow)
has a much higher redshift (z = 0.067). (Hale Observatories.)

Figure 7-11   Stephan's Quintet
Four galaxies in this cluster have nearly the same redshift
($z = 0.02$). But the remaining galaxy (indicated by an arrow)
has a much lower redshift ($z = 0.003$). (Kitt Peak National
Observatory.)

ered that the fifth galaxy (NGC 7320) has a much lower redshift than the other four. Is Stephan's Quintet really a quintet that violates the Hubble law? Or is it really a quartet consisting of a foreground galaxy that just happens to appear in the same part of the sky as a cluster of four remote galaxies? Although most astronomers would argue in favor of the traditional Hubble-law interpretation, the issue remains open.

In 1978, Dr. Arp published his observations of one of the most intriguing examples of a possible violation of the Hubble law. It seems that he has discovered a high-redshift, quasarlike object silhouetted *in front* of the low-redshift elliptical galaxy called NGC 1199. This case is shown in Figure 7-12. The galaxy has a redshift of $z = 0.009$, which corresponds to a speed of 2,600 kilometers per second and a Hubble distance of only 150 million light years. But the compact, quasarlike object has a redshift of $z = 0.044$, which corresponds to a speed of 13,300 kilometers per second. If the Hubble law is obeyed, the compact object should be at a distance of nearly 800 million light years. But it looks as if the compact object is in front of the galaxy! Traditional astronomers argue that "appearances can be deceiving." They feel that the Hubble law is obeyed and that Figure 7-12 really shows a remote quasarlike object that is *shining through* the nearby galaxy. Of course, Arp does not agree. Once again, the issue remains open.

It should be emphasized that the mavericks are in a distinct but vocal minority. Most astronomers feel that redshifts are good indicators of distance according to the Hubble law. They feel that all of Arp's cases amount to nothing more than chance alignments of foreground and background objects. They argue that, if you searched hard enough, you could find many of these coincidences. For example, Figure 7-13 shows the "double quasar" called 1548 + 114a and b. The brighter quasar has a redshift of $z = 0.436$, corresponding to a speed of 35 percent of the speed of light. According to the Hubble law, this quasar should be at a distance of 6 billion light years. The fainter quasar has a much larger redshift ($z = 1.901$), corresponding to a speed of 79 percent of the speed of light. According to the Hubble law, the dim quasar should be 14 billion light years away. But the two quasars appear side by side. "Pure chance," according to the traditionalists.

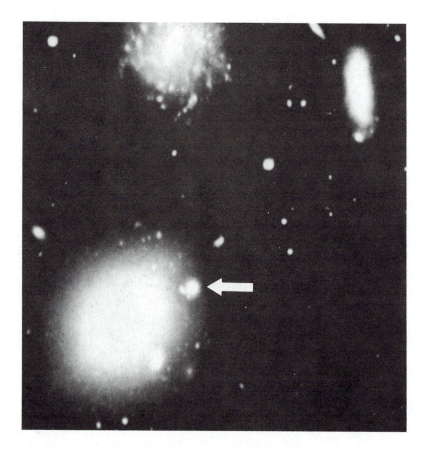

Figure 7-12   *A Quasar* in Front *of a Galaxy?*
*A compact, quasarlike object appears to be silhouetted in front of
the elliptical galaxy NGC 1199. The compact object has a high
redshift (z = 0.044) and the galaxy has a low redshift (z = 0.009).
If the compact object is really closer to us than the galaxy, this case
would constitute a major violation of the Hubble law. (Courtesy of
H. C. Arp; Hale Observatories.)*

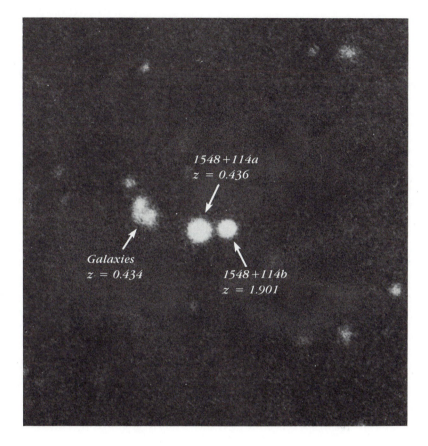

Figure 7-13   The Double Quasar 1548 + 114a and b
*Although these two quasars appear close together in the sky, they
have very different redshifts. According to the Hubble law, the
fainter quasar should be more than twice as far away from us as
the brighter member of the pair. Also notice that a faint cluster of
galaxies has almost exactly the same redshift as the brighter quasar.
This strongly suggests that the bright quasar is in fact at the
distance inferred from its redshift and the Hubble law. (Courtesy of
E. M. Burbidge and Harding E. Smith; Kitt Peak National
Observatory.)*

In support of the traditional viewpoint, several faint galaxies have been found near the "double quasar." These galaxies all have nearly the same redshift as the brighter (and, presumably, nearer) member of the pair. This strongly suggests that the quasar 1548 + 114a and the galaxies are all at the same distance. Evidently 1548 + 114a does obey the Hubble law. In addition, 1548 + 114a is a source of radio waves. It is called 4C11.50 in the *Fourth Cambridge Catalogue.* But the high-redshift quasar 1548 + 114b is *radio-quiet.* There are many quasars all across the sky that do not emit radio waves. These radio-quiet quasars may be far more plentiful than we suspect. (Of course, they are hard to find because they aren't emitting radio waves to distinguish themselves from ordinary stars.) If radio-quiet quasars are indeed very numerous, then coincidences like the chance alignment between 1548 + 114a and 1548 + 114b should not raise any eyebrows.

Powerful evidence supporting the traditional approach to the redshift controversy came late in 1978, when Alan Stockton of the University of Hawaii published his observations of the regions surrounding 27 low-redshift quasars. Dr. Stockton wanted to see if any galaxies could be discovered in the vicinity of these quasars, just like the galaxies huddled close to 1548 + 114a in Figure 7-13. Indeed there were. In eight cases, the redshifts of the galaxies were virtually the same as the redshifts of the quasars around which they are clustered. Several of these cases are shown in Figure 7-14.

Stockton's observations provide strong support in favor of the idea that quasars obey the Hubble law. He has succeeded in finding quasars embedded in remote clusters of galaxies. Because the Hubble law works for ordinary galaxies, and because each of Stockton's eight quasars has nearly the same redshift as the galaxies that surround it, we must conclude that the redshifts of the quasars themselves obey the Hubble law.

These observations are crucial. They may be the deciding factor in the long-standing redshift debate between the traditionalists and the mavericks. Hubble discovered cepheid variables scattered around three enigmatic "spiral nebulas" and thereby settled the Shapley-Curtis debate. Perhaps the faint galaxies that Stockton has found scattered around eight enigmatic quasars will finally settle the redshift debate.

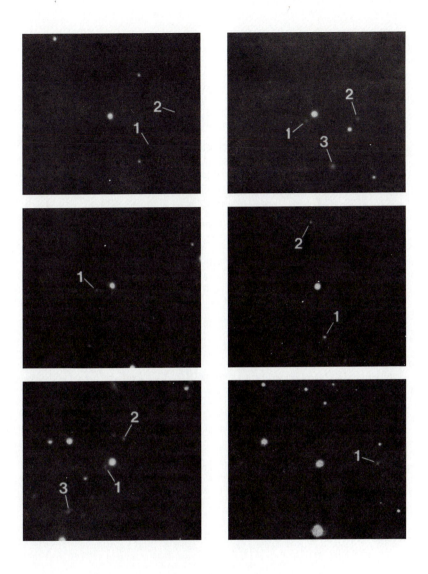

Figure 7-14   Quasars in Remote Clusters of Galaxies
A quasar is located at the center of each of the above photographs.
Very distant galaxies are faintly seen (identified by numbers) near
each of the quasars. In each case the redshift of the quasar is
virtually the same as the redshifts of the galaxies that surround it.
(Courtesy of Alan Stockton.)

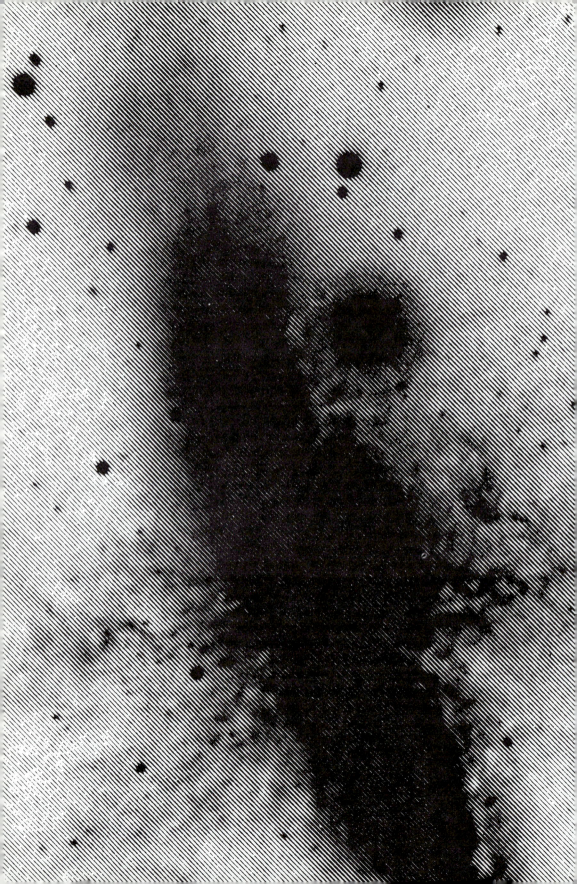

8

Exploding Galaxies and Supermassive Black Holes

Cygnus A was one of the first radio sources to be discovered. Its radio emissions were picked up with makeshift, backyard equipment while radio astronomy was still in its infancy. At that time, many scientists felt that radio astronomy would probably remain an amusing, yet interesting, sideline. They speculated that radio telescopes would forever be relegated to second-class citizenship: poor stepchildren to the magnificent optical machines, such as the colossal 200-inch telescope on Palomar Mountain, that had recently been put into operation. They could not be more wrong.

The breakthrough came in 1954, when Cygnus A was identified with a very remote peculiar galaxy shown in Figure 7-1. Obviously, those astronomers with all their wires and amplifiers had stumbled upon an enormous source of energy. Cygnus A was shining more brilliantly than anything that astronomers had ever seen. And most of that prodigious energy output was at radio wavelengths. Anyone with any doubts about the future of radio astronomy immediately did an about-face.

As the quality of radio telescopes improved, it was discovered that virtually all of the radio emission from Cygnus A was *not* coming from the peculiar galaxy itself. Instead, the radio waves were coming from two "blobs" located on either side of the peculiar galaxy.

A radio map of Cygnus A is shown in Figure 8-1. Radio astronomers often display their observations with contour maps of this type. Just as contour lines on a surveyor's map indicate the height of the land, the contour lines on a radio astronomer's map indicate the intensity of radio emissions. Notice that the radio emissions from Cygnus A are concentrated in two "blobs." The peculiar galaxy is located exactly in between the two "blobs." Each of these "blobs" is the size of a galaxy, and each shines with the luminosity of a trillion suns! And *all* that energy output is entirely at radio wavelengths. Even the longest photographic exposures with the 200-inch telescope fail to reveal anything but empty sky at the positions of the "blobs."

While astronomers in the Northern Hemisphere were puzzling over Cygnus A, their colleagues in Australia had discovered radio waves coming from a galaxy in the southern constellation of Centaurus. The source is called Centaurus A and is associated with the strange galaxy NGC 5128. A photograph of NGC 5128 is shown in

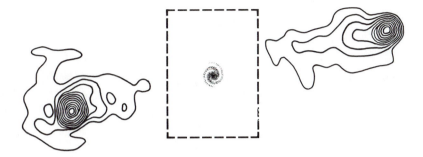

Figure 8-1 A Radio Map of Cygnus A (also called 3C405)
This contour map of radio emissions from Cygnus A shows that nearly all of the radio waves come from two "blobs" located on either side of a peculiar galaxy. The dashed rectangle indicates the area of the sky covered in Figure 7-1.

Figure 8-2, and the corresponding radio map is given in Figure 8-3. Again notice that most of the radio emissions are coming from two "blobs" on either side of the galaxy. In addition, two more "blobs" are located in the galaxy itself. Whatever produced the first pair of "blobs" seems to be making a second pair.

In spite of its dramatic appearance, Centaurus A is not an exceptionally powerful radio source. It is a nearby galaxy, only 10 million light years away. For comparison, Cygnus A is 1 billion light years away and is 1,000 times as luminous as Centaurus A. Nevertheless, Centaurus A covers a much larger portion of the sky than most other radio sources. It measures 10 degrees from end to end. That is the same as twenty full moons placed side by side. Because of its proximity and large size, astronomers succeeded in mapping Centaurus A with the crude radio telescopes of the late 1940s. Comparable detailed studies of Cygnus A (which covers a much smaller portion of the sky because of its great distance) had to wait until the 1950s for the development of radio interferometers which gave a much sharper radio picture.

Figure 8-2   The Peculiar Galaxy NGC 5128
*This exotic galaxy is a powerful source of radio waves and X rays.
It is located about 10 million light years away, in the direction of
the constellation of Centaurus. The galaxy's distorted appearance
suggests that an explosive event or process may have occurred at
the galaxy's center. (Hale Observatories.)*

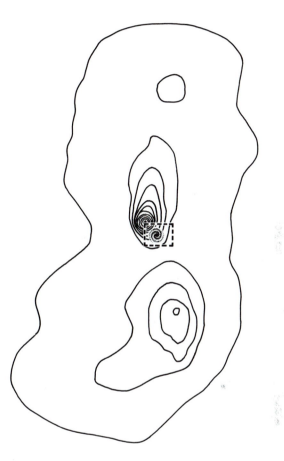

Figure 8-3   A Radio Map of Centaurus A

This contour map of radio emissions from Centaurus A shows that most of the radio waves come from two "blobs" located on either side of the galaxy. A second pair of "blobs" is located inside the galaxy itself. The dashed rectangle indicates the area of the sky covered in Figure 8-2.

As the quality of radio telescopes improved, astronomers began finding many of these *double radio sources* all across the sky. Like Cygnus A, they all consist of two locations (most astronomers prefer to call them "radio lobes" instead of "blobs") from which vast quantities of radio waves are pouring. Indeed, Cygnus A is the prototype. Some of these double radio sources are fat and some are skinny. Some are big and some are small. But their radio maps all bear a striking similarity to that of Cygnus A, shown in Figure 8-1. In many cases, a peculiar galaxy is located in between the two radio lobes.

Double radio sources posed a huge dilemma for astrophysicists, and it persists to this day. What causes the two "blobs"? Could it be an explosion, as suggested by the distorted appearance of the strange galaxies that we often find in between the two radio lobes? And what kind of process could eject two "blobs of something" in exactly opposite directions from the central galaxy?

As observations progressed, it became clear that the radio emission from these double sources is caused by high-speed electrons spiraling around a magnetic field. This is called *synchrotron radiation.* Whenever electrons move through a magnetic field, they are deflected by the magnetic field and forced to follow curved paths. As the electrons spiral around the magnetic field, they emit radio waves, as diagrammed in Figure 8-4.

We now know that synchrotron radiation is one of the most important processes by which astronomical objects emit radio waves. All you need are high-speed electrons (traveling near the speed of light) and some magnetic field. Evidently, in many astronomical situations, both are quite common.

But these insights about synchrotron radiation do not shed much light on the dilemma of Cygnus A and its cousins. There must be a continuous supply of high-speed electrons streaming into the radio lobes from the peculiar galaxy. What produces all these electrons? And how are they channeled into the lobes?

The discovery of double radio sources was one of our first clues that incredibly energetic processes are associated with some galaxies and galaxylike objects. The next major clue came with the discovery of quasars in the early 1960s. As I explained in the previous chapter, quasars shine with luminosities that are hundreds of times as great as those of ordinary galaxies.

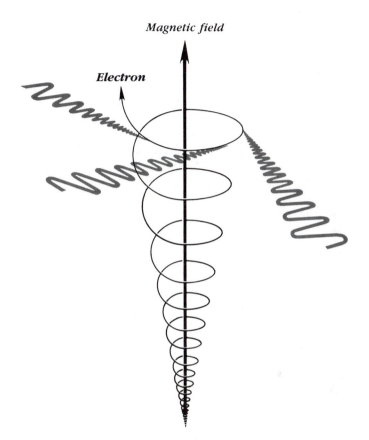

Figure 8-4   Synchrotron Radiation
High-speed electrons emit radio waves as they spiral around a magnetic field. Radiation produced in this fashion is called synchrotron radiation.

When quasars and double radio sources were first discovered, there seemed to be an enormous gap between ordinary galaxies and these exotic objects. But careful reexamination of galaxies that had been thought to be "ordinary" soon revealed surprising evidence that seems to bridge the chasm.

The 82nd object in Messier's catalog is one member in a small, nearby cluster of galaxies in the constellation of Ursa Major. Detailed photographs (see Figure 8-5) and spectroscopic studies by Allan Sandage and Roger Lynds in the 1960s revealed enormous amounts of gas erupting from the galaxy's nucleus. Filaments and plumes reach out to distances over 10,000 light years and are expanding with speeds of up to 1,000 kilometers per second. Corroborating observations made in the late 1970s all point to a colossal explosion that must have occurred at the galaxy's nucleus roughly 2 million years ago. Obviously, some galaxies are not the peaceful places they were thought to be. But, then, what could possibly make a galaxy blow up? Enormous amounts of energy are obviously involved.

Another good example of one of these exploding galaxies is NGC 1275 in the constellation of Perseus. It is the brightest galaxy in the Perseus cluster and is a powerful source of radio waves (called 3C84 or Perseus A by radio astronomers). Photographs such as Figure 8-6 show huge filaments of gas extending many tens of thousands of light years from the galaxy's nucleus. Some of this gas is being expelled from the galaxy at speeds of 3,000 kilometers per second. These eruptive phenomena are clear evidence of an explosion. Indeed, NGC 1275 looks more like the Crab Nebula than like a galaxy!

In 1943, the American astronomer Carl Seyfert drew attention to a dozen galaxies that seemed somewhat unusual. The spectra of these galaxies display strong emission lines. And photographs of these galaxies always show an unusually bright, almost starlike, nucleus. Of course, ordinary galaxies have only absorption lines in their spectra, and their nuclei are dimmer and blend in with the surrounding stars.

The exploding galaxy NGC 1275 is one of these *Seyfert galaxies*. So is M77, shown in Figure 8-7. And so is NGC 4151, shown in Figure 8-8. Actually, we now realize that roughly 10 percent of the brightest galaxies in the sky are Seyfert galaxies.

Figure 8-5 *The Exploding Galaxy M82 (also called NGC 3034 and 3C231)*
Huge filaments of gas erupting out of this galaxy are evidence of a colossal explosion that occurred roughly 2 million years ago at the galaxy's center. This galaxy is also a powerful source of radio waves, X rays, and infrared radiation. (Hale Observatories.)

165

**Figure 8-6 The Exploding Galaxy NGC 1275
(also called 3C84)**
*This remarkable photograph shows filaments of gas erupting from
the center of a very distorted galaxy. This galaxy is also a powerful
source of X rays and radio waves. (Courtesy of Roger Lynds; Kitt
Peak National Observatory.)*

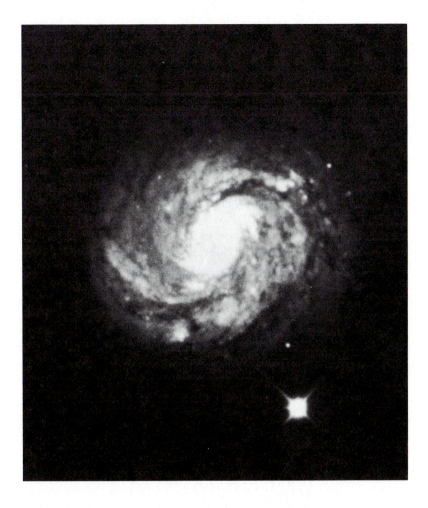

Figure 8-7 The Seyfert Galaxy M77 (also called NGC 1068)
*This galaxy has an unusually bright, almost starlike, nucleus. It is
an extremely powerful source of infrared radiation. Infrared
radiation from the galaxy's nucleus varies from week to week.
(Lick Observatory.)*

167

Recent infrared observations of M77 by Frank Low and G. H. Rieke of the University of Arizona revealed that this Seyfert galaxy is an extremely powerful source of infrared radiation. In fact, the amount of infrared radiation coming from the center of M77 equals 100 billion suns! Even more astounding, Rieke and Low have observed substantial fluctuations at infrared wavelengths. Over a period of a few months, the nucleus of M77 switches on and off a power output equivalent to the total luminosity of our entire Galaxy!

In addition to having an extremely bright nucleus, M77 also exhibits explosive phenomena. Merle Walker has found evidence for several huge clouds of gas erupting from the galaxy's nucleus. Each cloud probably contains at least a million solar masses of matter and measures nearly 1,000 light years in diameter. Speeds of up to 600 kilometers per second have been deduced from the Doppler shifts of the spectral lines of these clouds.

These eruptive phenomena are child's play in comparison with NGC 4151. Kurt Anderson and Robert Kraft of Lick Observatory have found evidence for three huge shells of gas being expelled from the galaxy's nucleus. They estimate that roughly 100 solar masses of matter per year are ejected from the center of NGC 4151, perhaps in a series of sporadic outbursts. Like M77, short-term variability is also observed in the nucleus of NGC 4151.

Take a good, hard look at NGC 4151 in Figure 8-8. This is a typical Seyfert galaxy. It has an extremely bright nucleus that exhibits explosive phenomena and short-term variability. But it is fairly nearby. The distance to NGC 4151 is only 40 million light years.

Suppose NGC 4151 were much farther away. Suppose that NGC 4151 were over a billion light years away. At that distance, it would be impossible to see any of the nebulosity or stars or spiral arms associated with the galaxy. You would *only* see the bright, star-like nucleus. And that starlike nucleus exhibits short-term variability; it flickers in brightness once a year or faster. And the spectrum of this starlike object contains strong emission lines, just like the spectrum of 3C273 shown in Figure 7-4. Obviously, *NGC 4151 would be mistaken for a quasar!*

Perhaps we have begun to realize that there is a continuous family of objects that range all the way from ordinary galaxies like our own up to the superluminous quasars. At one end of this family,

Figure 8-8   The Seyfert Galaxy NGC 4151
*The nucleus of this galaxy is so bright that if it were very far away,
it would probably be mistaken for a quasar. There is strong
evidence for the violent ejection of gas from the nucleus of this
galaxy. (Hale Observatories.)*

we have normal galaxies. In several cases, there is evidence for low-level activity and mild outbursts. For example, as we saw in Chapter 4, a weak explosive event may have occurred at the center of our Galaxy about 10 million years ago. There are also several bright infrared sources surrounding our galactic nucleus. Nevertheless, our Galaxy is quite ordinary and normal.

On the next rung up the ladder, we find active galaxies such as M82 and NGC 5128. These galaxies usually have a very distorted appearance and show clear evidence of explosive ejection of gas from their nuclei. They are also more powerful sources of radio waves, X rays, and infrared radiation than ordinary galaxies. In many respects, the phenomena associated with galaxies such as M82 are just an exaggerated version of the low-level activity going on in our own Galaxy. Perhaps if the activity surrounding Sagittarius A were increased tenfold, our Galaxy would begin looking like M82.

Taking yet another step up the ladder, we find Seyfert galaxies, such as M77 and NGC 4151. These galaxies display a higher level of activity than do objects such as M82. Indeed, in many respects, the properties of Seyfert galaxies overlap the properties of quasars. Perhaps a Seyfert galaxy is simply a nearby quasar. Of course, the superluminous quasars are at the top of the ladder.

In many respects, the whole business is like an expedition to the Olduvai Gorge in East Africa in search of prehuman skeletal remains. Occasionally we find a skull, a femur, or a tooth. Differences and similarities are noted. As data accumulate, connections and relationships start to emerge. Of course, there are still many missing links. But eventually the pieces begin to fit together.

Anthropologists delve into the past by excavating buried strata. The deeper they dig, the more ancient the bones that they find. Astronomers probe the past by gazing out into space. The photograph of a galaxy 100 million light years away shows how that galaxy looked 100 million years ago. The farther out we look, the more ancient the objects that we see. Remote quasars whose light takes billions of years to reach us are the most archaic objects we have ever seen.

Anthropologists often conclude that creatures whose bones are found in upper, recently buried strata must have evolved from animals whose remains are found in deeper, more ancient strata. In the same way, it seems reasonable to suppose that ancient, remote objects we see in the sky are the ancestors of the nearby objects. Perhaps quasars evolve into Seyfert galaxies. As Seyfert galaxies get older, they become more subdued and turn into active galaxies. As active galaxies age, their energy wanes and they become ordinary galaxies. The occasional weak outbursts that we see in ordinary galaxies like our own are perhaps dim, half-hearted reminders of a vigorous and violent youth many billions of years ago.

Of course, only a small percentage of present-day galaxies could have evolved from quasars. For example, it is inconceivable that dwarf galaxies — which are so numerous — were once quasars. And in addition, the actual path of evolution may be far more complicated than simply quasar-to-Seyfert-to-active-to-ordinary. After all, anthropologists have unearthed the bones of many apelike, prehuman creatures that are not directly related to us. For example, Australopithecus was once believed to be our directly related ancestor but is now recognized as one of many evolutionary dead ends. Nevertheless, evidence is growing that quasars are the ancestors of at least *some* galaxies. We do indeed see quasarlike activity at the centers of some galaxies. Surely it would be extremely enlightening if we could find evidence for galaxies surrounding some quasars.

There is a class of objects in the sky called *BL Lacertae objects.* They are named after their prototype, BL Lacertae, shown in Figure 8-9. Like the first-discovered quasars, BL Lacertae was once believed to be an unusual, variable, nearby star in the constellation of Lacerta. We now know that it is an extremely remote object located out among the galaxies. These BL Lacertae objects are powerful sources of radio waves and infrared radiation. Like true quasars, they have a distinct starlike appearance and exhibit short-term brightness fluctuations. But unlike quasars, the spectra of BL Lacertae objects do not exhibit any spectral lines. Their spectra are smooth and featureless. Obviously, it has been impossible to measure redshifts, and consequently BL Lacertae objects have been poorly understood.

Although some properties of BL Lacertae objects resemble quasars, other properties resemble Seyfert galaxies. For example, BL Lacertae itself seems to be surrounded by nebulosity. Some of this

Figure 8-9   BL Lacertae
This is the prototype of a class of starlike objects that astronomers call BL Lacertae objects. They are like quasars, except that their spectra do not exhibit spectral lines. Notice the "fuzz" that surrounds BL Lacertae. (Courtesy of Tom Kinman; Kitt Peak National Observatory.)

"fuzz" is easily seen in Figure 8-9. In this respect, BL Lacertae resembles a distant Seyfert galaxy whose outlying stars are barely discernible in the glare of the bright galactic nucleus.

During the summer of 1977, a team of astronomers at Lick Observatory succeeded in obtaining a spectrum of the "fuzz" that surrounds BL Lacertae. Joseph S. Miller, Howard B. French, and Steven A. Hawley evaporated a small aluminum dot onto a glass microscope slide. They then placed the glass slide at the focus of the observatory's 120-inch Shane reflector, in between the telescope and their spectroscopic equipment. Because of the dot, they were able to block out the glaring light from the nucleus of BL Lacertae. Fainter light from the surrounding "fuzz" streamed past the edges of the dot and continued on into the spectroscopic equipment.

The spectrum of the "fuzz" around BL Lacertae is shown in Figure 8-10. Notice the striking similarity to the spectrum of M32, a small, nearby elliptical galaxy. (M32 is a companion to M31 and can be seen alongside the huge spiral galaxy in Figure 2-4.) Because of this extraordinary similarity, Miller and his colleagues conclude that the bright, variable nucleus of BL Lacertae is embedded in an otherwise ordinary elliptical galaxy of stars. The redshift of this "underlying galaxy" is $z = 0.07$, which corresponds to a distance slightly over 1 billion light years from Earth. These observations confirmed similar work by John Beverley Oke and James Gunn at the 200-inch telescope on Palomar Mountain.

The discovery of a galaxy surrounding the quasarlike nucleus of BL Lacertae lends strong credence to the idea that quasars are simply the superluminous centers of young galaxies that are experiencing a violent and energetic youth. Perhaps we might therefore gain further insight into this issue by observing nearby clusters of galaxies and examining some of their biggest, brightest members.

The Virgo cluster is the nearest rich cluster of galaxies in the sky. As mentioned in Chapter 5 (see Figure 5-11), the Virgo cluster is only 60 million light years away. Near the approximate center of this sprawling cluster is the supergiant elliptical galaxy called M87. An exceptionally fine photograph of M87 is shown in Figure 8-11.

In 1918, H. D. Curtis discovered a luminous "jet" surging up out of the nucleus of M87. This jet shows up only on short time

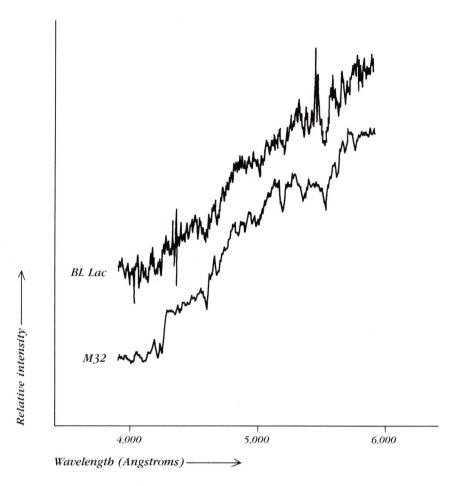

***Figure 8-10 The Spectrum of the Galaxy Surrounding BL
Lacertae***
*The upper curve is the spectrum of the "fuzz" that surrounds BL
Lacertae. The lower curve is the spectrum of the ordinary elliptical
galaxy M32. The striking similarity between the two curves is
evidence for the fact that BL Lacertae objects are embedded in
galaxies. (Adapted from J. S. Miller, H. B. French, and S. A. Hawley.)*

exposures such as those in Figure 8-12 (with long time exposures, such as Figure 8-11, light from all the surrounding stars completely swamps the image of the jet). The lower view in Figure 8-12 is a computer-enhanced version made by Jean J. Lorre at Caltech's Jet Propulsion Laboratory from several photographs supplied by H. C. Arp. The jet clearly consists of a series of bright knots in perfect alignment.

M87 is definitely not an ordinary, run-of-the-mill galaxy. It has a bright, almost starlike, nucleus that shows up well in Figure 8-12. Indeed, the bright nucleus along with the jet is suspiciously reminiscent of the quasar 3C273 shown in Figure 7-3. The jet in M87 is 6,500 light years long and is a powerful source of synchrotron radiation.

In 1978, two teams of astronomers announced the results of their detailed observations of M87. One team, headed by Peter Young at Caltech, made very precise measurements of the brightness all across the galaxy. The second team, headed by Wallace Sargent, also at Caltech, made equally precise spectroscopic observations all across the galaxy. These two sets of observations gave an accurate picture of how stars are distributed in M87 (from brightness measurements) and how fast they are all moving (from spectrosopic measurements).

Ordinary elliptical galaxies simply consist of a large number of stars orbiting their common center. Using classical Newtonian mechanics, it is possible to calculate how the stars are distributed and how fast they should be moving. In many respects, the calculations are analogous to the kind of computations that give the orbits of the planets about the sun. For example, just as Mercury orbits the sun faster than Pluto does, stars near the center of a galaxy must be moving more rapidly than those that leisurely orbit the periphery. Calculations of this type were first done in 1966 by Ivan R. King at the University of California at Berkeley. King's theoretical models give us an understanding of how stars should be distributed inside an ordinary elliptical galaxy.

To everyone's surprise, the new data about M87 did not fit the standard models of King. For example, Figure 8-13 shows the brightness across M87, measured outward from the galaxy's center. Each

**Figure 8-11   The Supergiant Elliptical Galaxy M87 (also
called NGC 4486, Virgo A, and 3C274)**
*This enormous galaxy in the Virgo cluster is a powerful source of
radio waves and X rays. This long time exposure reveals hundreds
of faint globular clusters that surround the galaxy like a swarm of
bees. This photograph was taken with the 4-meter telescope at Kitt
Peak. The view in Figure 2-11 was taken with the 5-meter telescope
at Palomar. (Kitt Peak National Observatory.)*

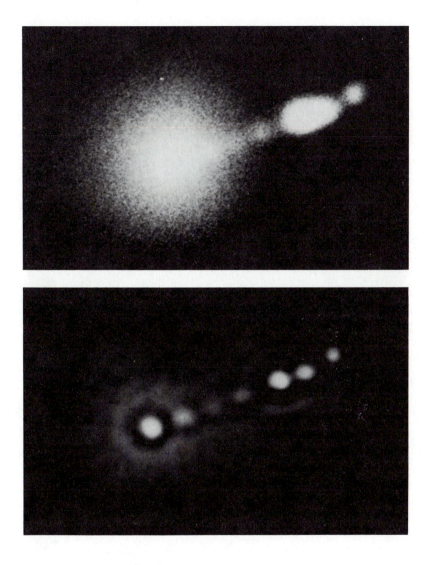

Figure 8-12   The "Jet" in M87
A short time exposure reveals a luminous jet surging out of the nucleus of M87. The upper view is a single, ordinary photograph of the jet. The lower view is a computer-enhanced exposure made from several photographic plates. (Courtesy of H. C. Arp and J. J. Lorre; Hale Observatories.)

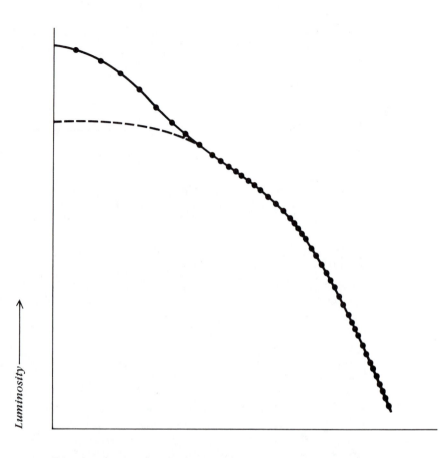

Distance from galaxy's center ⟶

Figure 8-13   The Luminosity Profile of M87
This graph shows the brightness across M87, measured outward
from the galaxy's center. Each dot represents an individual
measurement. The dashed curve is the standard model according to
King. The solid curve (which fits the data!) is the standard model
plus a supermassive black hole at the galaxy's center. (Adapted from
P. Young and colleagues.)

dot represents a measurement by Dr. Young and his colleagues. In the outer portions of the galaxy, the data agree very well with King's theoretical calculations (dashed curve). But in the central regions of the galaxy, the data points fall far above the dashed curve based on King's model. This strongly suggests that a very massive object at the core of M87 is causing many stars to be huddled close to the galaxy's nucleus. In fact, the Caltech astronomers found that they could account for the brightness across M87 by adding a *supermassive black hole* to King's model. By assuming that a very massive nonluminous object resides at the core of M87, it is possible to explain completely the brightness measurements across the galaxy. In a similar fashion, the spectroscopic measurements across the galaxy also strongly indicate the presence of an extremely massive black hole. This is the only way that Dr. Sargent and his colleagues could explain the unusually large spread in stellar velocities they found by examining the spectrum of starlight from M87's central regions. The two teams of astronomers conclude that the mass of this huge black hole is about 5 billion suns.

The concept of a black hole follows logically from the General Theory of Relativity, which is our best theory about how gravity works. A black hole represents the complete triumph of gravity over all other forces in nature. For example, when a massive star has used up all of its internal sources of energy, the enormous weight of trillions upon trillions of tons of burned-out stellar matter pressing inward causes the star to contract. As the contraction continues, gravity around the star becomes stronger and stronger. This causes the geometry of space and time around the star to become increasingly curved. If the dying star is massive enough (that is, at least three times as massive as the sun), then nothing can stop the star from completely collapsing in upon itself. Quite simply, no forces in nature are strong enough to support a massive dead star against its own weight. As the collapse proceeds to its inevitable end, space and time fold over themselves and the star disappears from the universe. All that remains is a hole in the fabric of space and time. The gravity of this hole is so strong that nothing — not even light — can escape.

It is reasonable to suppose that there are many black holes scattered around our Galaxy. These black holes are simply the corpses of burned-out massive stars. Several of these ordinary black holes have already been identified.*

The black hole that may reside at the core of M87 is not one of these ordinary stellar black holes. The most massive stars in the sky are only 50 times as massive as our sun. Consequently, the corpse of one of these stars can contain, at most, only 50 solar masses of completely collapsed matter. The mass of the huge black hole in M87 is many hundreds of millions of times as great.

No one really knows how one of these supermassive black holes might form. Some astronomers argue that there is so much congestion at the center of a large galaxy that ordinary black holes collide and gobble up each other. When two black holes swallow each other, they create a bigger black hole. Gradually, over the years, the hole grows to enormous proportions.

Another possibility is that supermassive black holes were created during the Big Bang. The birth of the universe must have been exceedingly violent. Perhaps small "lumps" in the newborn universe were crushed into many tiny black holes, as first suggested by Stephen W. Hawking of Cambridge University. Because of the tremendous density during the first few seconds, these black holes could have grown to considerable proportions — literally having been force fed — before the universe had expanded significantly. These supermassive black holes then became the "seeds" around which galaxies and clusters of galaxies eventually formed.

M87 was not the first case in which astronomers had discussed the possibility of supermassive black holes. In the late 1960s, in an attempt to account for activity at the center of our own Galaxy, Donald Lynden-Bell proposed that a supermassive black hole might be located at or near Sagittarius A. Lynden-Bell argued that gas should be captured into orbit about the hole. Consequently, a vast disk of material, called an *accretion disk,* circles the hole, like a giant version of the rings around Saturn. As the gases spiral in toward the hole, they are heated by friction to high temperatures. Before taking the final plunge into the hole, the hot gases in this swirling whirlpool are

---

*For a more detailed discussion of black holes, you are referred to *The Cosmic Frontiers of General Relativity* (Little, Brown, Boston, 1977) by the author of this book.

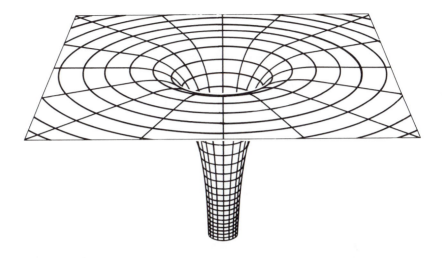

Figure 8-14 A Black Hole
A black hole is literally a hole in space and time. The gravitational field of the hole is so strong that nothing —not even light —can escape.

radiating so fiercely that they supply all the energy needed to explain all the activity we observe at our Galaxy's core.

Supermassive black holes have also been invoked to explain extraordinary activity at the centers of many other galaxies. For example, in 1976, A. C. Fabian and his colleagues at Cambridge argued that matter falling into a huge black hole at the center of NGC 5128 would account for all the properties of this intriguing galaxy. They estimated the mass of the hole to be 10 million suns. This idea received further support in 1978, when astronomers at Harvard discovered a variable X-ray source at the center of NGC 5128. This X-ray source is very compact and varies in brightness by 25 percent in a few hours. The X-ray source cannot therefore be larger than a few light hours across. That is roughly the same dimension as the inner portions of an accretion disk surrounding the supermassive black hole.

Incidentally, a 10-million-solar-mass black hole has a diameter of 3 light minutes (60 million kilometers). And a 5-billion-solar-mass black hole, such as the one that resides at the core of M87, has a diameter of 28 light hours (30 billion kilometers). For comparison, the diameter of Pluto's orbit is 11 light hours (12 billion kilometers). Obviously these supermassive black holes are comparable in size to our solar system.

There are two reasons why supermassive black holes are good candidates for explaining the enormous energy output from exploding galaxies and quasars. First, supermassive black holes are small. They are typically a light day in diameter, or smaller. This small size permits the rapid variability in brightness that is commonly seen in many quasars and related objects. Second, the enormous gravitational field associated with one of these holes is, in principle, a vast source of energy. The real issue, therefore, becomes one of explaining in detail how it is reasonably possible to extract some of the vast quantities of energy that are tied up in the powerful gravitational field of one of these holes.

The first black hole models of active galaxies, such as those proposed by Lynden-Bell, simply involved the formation of an accretion disk surrounding the hole. In recent years, however, two important complications have been added to black hole calculations, by R. V. E. Lovelace at Cornell and R. D. Blandford at Cambridge. These two embellishments, which seem to point the way to a complete understanding of quasars, are rotation and a magnetic field.

We would naturally expect the supermassive black hole at the center of a galaxy or quasar to be rotating. After all, galaxies themselves rotate. But rotating black holes have some strange properties. For example, near the hole, space and time are literally dragged around the hole. It is impossible to stand alongside a rotating black hole without getting pulled around the hole.

We would also expect that the gases captured into the accretion disk about the hole bring along some of the galaxy's magnetic field. We know that most galaxies and quasars must be threaded by magnetic fields. After all, only with a magnetic field can high-speed electrons produce the synchrotron radiation that we so readily observe. But as the matter in the accretion disk spirals in toward the hole, the magnetic field carried by the gases becomes very concen-

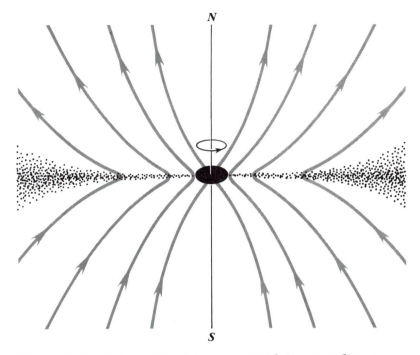

***Figure 8-15   A Magnetized Accretion Disk Surrounding a
Supermassive Black Hole***
*Gases captured by a rotating black hole form a disk in the plane of
the hole's equator. These gases also carry some of the galaxy's
overall magnetic field. As the gases spiral inward toward the hole,
the magnetic field becomes extremely concentrated and is aligned
parallel to the axis of rotation of the hole.*

trated. Lovelace and Blandford have therefore argued that the *rotating supermassive black hole must be surrounded by a magnetized
accretion disk.* The arrangement of the hole, the accretion disk, and
the magnetic field are sketched in Figure 8-15.

The rotation of the hole tries to pull the inner portions of the
magnetic field around the hole at a very high speed. But, of course,
these portions of the magnetic field near the hole are firmly attached

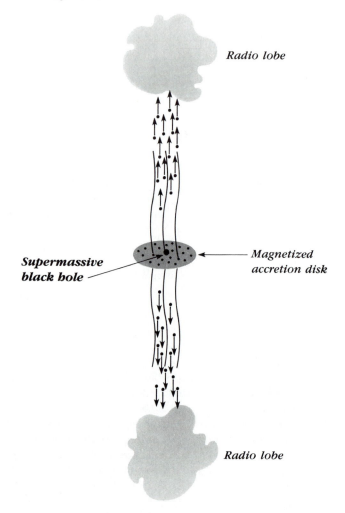

Radio lobe

**Supermassive
black hole**

*Magnetized
accretion disk*

Radio lobe

Figure 8-16   The Black Hole Dynamo
*The dragging of the magnetic field around the rotating black hole
produces a powerful electric field. This field is so strong that vast
numbers of electrons and positrons are created around the hole. The
electric field expels the electrons along two beams, guided by the
magnetic field.*

to the rest of the magnetic field at much greater distances. This is important, because nothing can go faster than the speed of light. Although the rotating black hole tries to pull the inner portions of the magnetic field around the hole at a high speed, the outer regions of the magnetic field simply cannot keep up the pace. The outer regions of the magnetic field far from the hole would have to move faster than light to keep up with the rapidly rotating inner regions. This is flatly impossible. Consequently, the magnetic field nearest the hole *slips* around the hole. These inner portions of the magnetic field are dragged around the hole at a speed less than that at which the hole itself is rotating.

In a major theoretical breakthrough in 1977, Blandford and R. L. Znajek showed that this slippage or dragging of the magnetic field around a rotating black hole generates a powerful electric field. The entire situation is literally a colossal dynamo that produces an extremely intense electric field that dominates space for thousands of light years around the hole. Indeed, the intensity of this field is so strong that pairs of electrons and positrons are created in huge amounts above the north and south magnetic poles. Because of the electric field, the electrons are vigorously propelled outward from the hole along the magnetic axes, as diagrammed in Figure 8-16.

This dynamo model is extremely promising. All you need is a rotating supermassive black hole and an accretion disk that carries a magnetic field. From this configuration, you get two powerful beams of high-speed electrons gushing continuously out of the north and south magnetic poles. It finally looks as though we have an explanation for all those double radio sources across the sky. It seems reasonable to suppose that a few variations on this basic theme will produce a detailed understanding of quasars and Seyfert galaxies. Of course, the astounding implication is that supermassive black holes must be exceedingly common, perhaps lurking at the centers of most major galaxies in the universe.

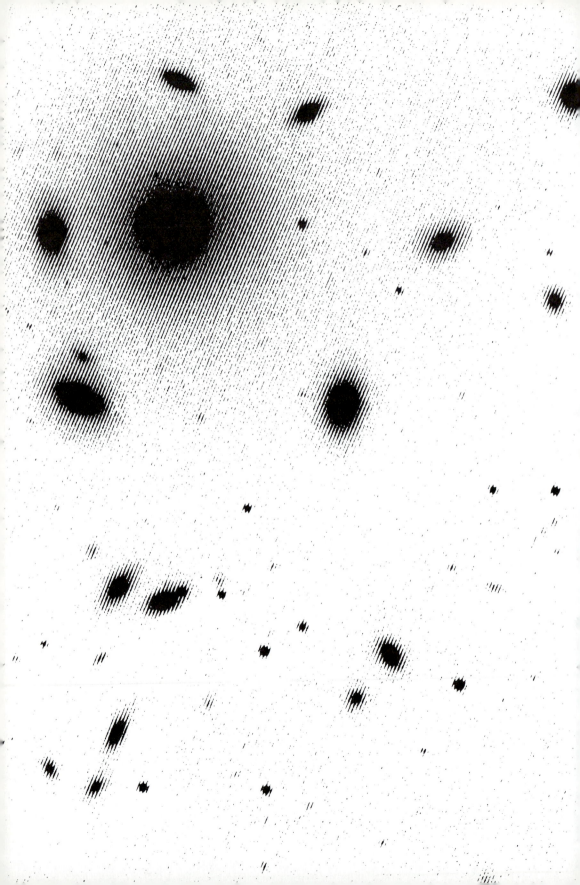

9

Relativistic Cosmologies

Silently, effortlessly, vast clusters of galaxies glide across space, coasting farther and farther apart. No force propels these colossi in their blind and futile plunge toward the nonexistent edge of the cosmos. All their impetus was imparted in one inconceivable blow dealt during the creation of the universe itself.

We live in an expanding universe. The distances between widely separated clusters are increasing. And the rate at which these clusters are moving apart is proportional to their separation. This is what "expansion" means. This is the simplest and most straightforward interpretation of the Hubble law.

But surely this cosmic expansion must be slowing down. All those galaxies scattered across space are exerting gravitational forces on each other. The mutual gravitational attraction among all the galaxies must be decelerating the frantic pace of universal expansion.

Will the expansion of the universe ever stop? Or will the rate of deceleration be too small to effectively inhibit the headlong rush of the cosmos perpetually toward the infinitely distant reaches of space? To answer these questions, to discover the ultimate fate of the universe, we must first construct theoretical models of the universe. Because gravity is the force that affects the evolution of the cosmos, we must be sure to use our best theory of gravity in our computations. Our models will therefore be based on the General Theory of Relativity. And, finally, we must examine motions of the most distant galaxies. We must try to accurately measure the rate at which the expansion of the universe is decelerating. By matching this deceleration with one of our relativistic models, we can calculate the destiny of the cosmos. In this way, the stars reveal our most distant future.

From the viewpoint of general relativity, you never need to speak about the "force" of gravity. Instead, Einstein's theory explains that gravity curves space and time. The stronger the gravity, the greater the curvature of space-time. And, finally, objects such as planets and light rays simply move along the shortest paths in curved space-time. This is the true meaning of general relativity: matter tells space-time how to curve, and curved space-time tells matter how to behave.

There is a lot of matter in the universe. Obviously, all this matter must have an effect on the geometry of space-time. The universe as a whole must have a *shape* because of all the matter in the universe. And furthermore, this geometry must influence the behavior of the matter across the cosmos. Consequently, the shape of the universe must be intimately related to the ultimate future of the universe. A universe that expands forever must have a very different shape from a universe whose expansion eventually stops and contraction begins. But what do we really mean by the "shape" of the universe?

Imagine that you shine two powerful laser beams out into space. Now suppose that you align these two beams so that they start off perfectly parallel. Finally suppose that nothing gets in the way of these two beams; we can follow them for billions of light years across the universe, across the space whose curvature we wish to detect.

There are only three possibilities. First, we might find that our two beams of light remain perfectly parallel, even after traversing billions of light years. In this case, space is not curved. The universe has *zero curvature* and space is *flat.*

Alternatively, we might find that our two beams of light gradually converge. The two beams gradually get closer and closer together as they move across the universe. Indeed, the two beams might eventually intersect at some enormous distance from the earth. In this case, space is not flat. Just as lines of longitude on the earth's surface intersect at the poles, the geometry of the universe must be like the geometry of a sphere. We therefore say that the universe has *positive curvature* and space is *spherical.*

The third and final possibility is that the two parallel beams of light eventually diverge. The two beams gradually get farther and farther apart as they move across the universe. In this case, the universe must also be curved. But it must be curved in the opposite sense of the spherical case. We therefore say that the universe has *negative curvature.* In the same way that a sphere is a positively curved surface, a saddle is a good example of a negatively curved surface. Just as parallel lines drawn on a sphere always converge,

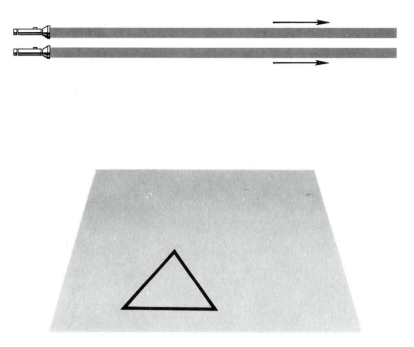

Figure 9-1 Flat Space
*If two parallel light beams remain parallel forever, then space has
zero curvature. The geometry of flat space is the three-dimensional
analogy of a two-dimensional plane. Geometrical properties of flat
space include the fact that the sum of the angles of a triangle is
exactly equal to 180 degrees.*

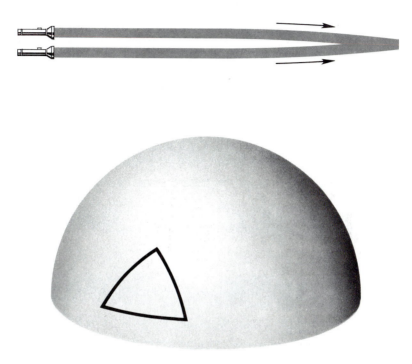

Figure 9-2 Spherical Space
If two parallel light beams eventually converge, then space has positive curvature. The geometry of positively curved space is the three-dimensional analogy of the two-dimensional surface of a sphere. Geometrical properties of spherical space include the fact that the sum of the angles of a triangle is always greater than 180 degrees.

parallel lines drawn on a saddle always diverge. Mathematicians say that saddle-shaped surfaces are "hyperbolic." Thus, in a negatively curved universe, we say that space is *hyperbolic*.

Each of these three cases corresponds to a different behavior and ultimate fate of the universe. To understand these different alternatives, imagine throwing a rock up into the air. Once again there are only three possibilities.

First of all, the rock may simply go up and come back down. In that case, the speed of the rock was *less* than the escape velocity from the earth.

Alternatively, you could throw the rock upward with a much greater speed, perhaps with the aid of a rocketship. If the speed of the rock *equals* the escape velocity from the earth, then the rock will never fall back. It will just manage to escape from the earth's gravitational pull.

The third possibility is that the rock is given a speed *greater* than the escape velocity from the earth. In this case, the rock has no trouble escaping the earth's gravity. Even after traveling an enormous distance, we find that the rock continues its journey away from the earth at a considerable speed.

By analogy with the rock, our question about the future of the universe becomes, Are clusters of galaxies rushing apart with speeds great enough to overcome their mutual gravitational attraction?

Cosmological models of the universe based on Einstein's General Theory of Relativity were first worked out by the Russian mathematician Alexandre Friedmann in 1922. Friedmann found that if there is enough matter in the universe to stop the expansion, then there is enough gravity to cause space to fold around on itself like a sphere. Thus, in a positively curved universe, the expansion eventually stops and contraction begins. In a spherical universe, the speeds of the galaxies are less than their mutual escape velocity.

If the galaxies are rushing apart with speeds that exactly equal their mutual escape velocity, then the universe will never collapse back upon itself. Of course, the expansion of the universe eventually slows to a snail's pace. But there is not quite enough gravity to cause the galaxies to completely stop in their tracks. Consequently, accord-

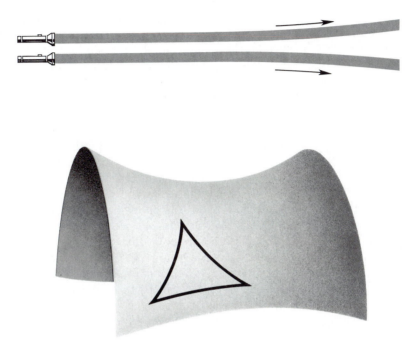

Figure 9-3   Hyperbolic Space
*If two parallel light beams eventually diverge, then space has
negative curvature. The geometry of negatively curved space is the
three-dimensional analogy of the two-dimensional surface of a
saddle. Geometrical properties of hyperbolic space include the fact
that the sum of the angles of a triangle is always less than 180
degrees.*

ing to Friedmann's models, there is not quite enough gravity to cause space to fold around on itself. This universe, therefore, has zero curvature. In a flat universe, the galaxies just barely manage to overcome their mutual gravitational attraction.

Finally, perhaps the mutual gravitational attraction between clusters of galaxies is so weak that the universal expansion will continue with vigor infinitely far into the future. According to general relativity, the shape of this universe must be hyperbolic. In a hyperbolic universe, clusters of galaxies will be rushing apart forever and ever.

In discussing the shape and fate of the universe, we are dealing with properties of the universe on an enormous scale. Individual galaxies are lost in the background as we gaze across distances far greater than the separations between superclusters. It is like looking at the palm of your hand. You know that you actually are made of an enormous number of tiny atoms, each of which consists of a dense nucleus orbited by electrons in otherwise empty space. But all this detail is lost as you run your fingers over your skin. In the same fashion, the universe is remarkably smooth and featureless in the enormous scale we are considering. In the scale of the Friedmann models, the universe is homogeneous and isotropic.

Because of the homogeneity of the universe on the largest scale, we can speak intelligently about the *average density* of the universe. We can act as though matter is uniformly spread across space, because individual details of one galaxy or another are totally insignificant. The advantage is that, by speaking about the average density, we can relate the amount of matter across space to the gravity that curves space. But we have just seen how the shape of the universe is intimately associated with the fate of the universe. Consequently, a measurement of the average density across the cosmos should enable us to predict the ultimate future.

The flat (zero curvature) model universe constitutes the dividing line between the spherical (positively curved) universe and the hyperbolic (negatively curved) universe. Consequently, the average density associated with a flat universe is the crucial quantity that differentiates the three cases. The average density of a flat universe is therefore called the *critical density*.

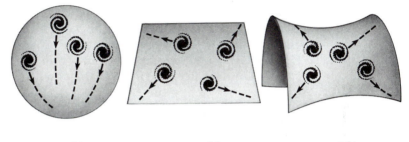

$$q_0 > {}^1/_2 \qquad\qquad q_0 = {}^1/_2 \qquad\qquad q_0 < {}^1/_2$$

Figure 9-4   Geometry and Destiny
*The ultimate fate of the universe is directly related to the geometry
of the universe. If space is spherical, universal expansion will
someday stop and contraction will begin. If space is flat, the
universe will just barely manage to expand forever. If space is
hyperbolic, the expansion will continue forever with vigor.*

The critical density is simply the average density our universe should have *if* it is really flat. With this density, the clusters of galaxies will just barely manage to overcome their mutual gravitational attraction. If there is a little more matter spread across space than the critical density, then there will be enough gravity to someday stop the expansion of the universe. If there is a little less matter spread across space than the critical density, then the universe will have no trouble expanding forever and ever.

Obviously, the critical density must be related directly to the rate at which the universe is expanding. After all, the critical density is the quantity that decides whether or not the expansion will continue forever. We measure the expansion rate of the universe simply by observing redshifts and distances of remote galaxies. The same data that were used to plot the Hubble law in Figure 5-6 give an expansion rate corresponding to a *critical density of three hydrogen atoms per 1,000 liters of space.* (That is the same as 6 kilograms of matter per billion billion cubic kilometers.)

The relationships between the various quantities that affect the future of the universe are summarized in the table on the facing page. A spherical universe is said to be "closed" in the same sense that the surface of a sphere closes around upon itself. In principle, if you traveled for a long time in a particular direction, you would eventually get back to your starting place, just like an adventurous sailor who circumnavigates the globe. Such a universe is finite. It does not extend forever and ever. Nevertheless, it does not have an edge or a boundary. Neither does it have a center. After all, you could travel forever around the earth and never get to the "edge" or the "center." In the same way, an astronaut could journey forever across a closed, spherical universe and never reach the "edge" or the "center."

In contrast to the spherical case, both flat and hyperbolic universes are infinite. They extend forever in all directions and are said to be "open." Actually, we might say that a flat universe is "just barely open," because this case is the dividing line between a closed, spherical universe and a wide-open, hyperbolic universe. Of course, these open universes do not have any edges or centers simply because they extend forever. Consequently, no matter which universe we live in, it is always meaningless to ask, "What is beyond the edge of the universe?" And because none of these universes has any edges or boun-

Geometry of space	Curvature of space	Average density throughout space	Deceleration parameter	Type of universe	Ultimate future of the universe
Spherical	Positive	Greater th. the criticai density	?ater ι ½	Closed	Eventual collapse
Flat	Zero	Exactly equal to the critical density	Exactly equal to ½	Flat	Perpetual expansion (just barely)
Hyperbolic	Negative	Less than the critical density	Between 0 and ½	Open	Perpetual expansion

daries, it is also meaningless to ask, "What is the universe expanding into?" Such questions are fundamentally nonsensical.

In trying to decide which universe we live in, we find that these relativistic Friedmann cosmologies really depend on only two important numbers. First of all, the cosmological models must depend on the rate at which the universe is expanding. The number that expresses this expansion rate is called the *Hubble constant* and is usually given the symbol H_0. Allan Sandage at Hale Observatories has spent many years carefully measuring the redshifts and distances of remote galaxies in an exhaustive effort to determine this expansion rate. His value for the Hubble constant is 17 kilometers per second per million light years. This simply means that, as you gaze out into space, you pick up 17 kilometers per second of universal expansion for each million light years. For example, a galaxy that is 100 million light years away should be receding from us at a speed of 1,700 kilometers per second.

The second important number expresses the rate at which the universal expansion is slowing down. This number is called the *deceleration parameter* and is usually given the symbol q_0. For a totally empty, wide-open, hyperbolic universe, the deceleration parameter equals zero ($q_0 = 0$). This is the extreme case. Quite simply, there is no matter or gravity to slow down the expansion at all.

For a flat universe, where there is just enough matter to ensure that the clusters of galaxies just manage to escape from each other, the deceleration parameter equals one-half ($q_0 = \frac{1}{2}$). And for closed, spherical universes, the deceleration parameter is greater than one-half. The greater the deceleration parameter, the sooner expansion stops and collapse begins.

The history of the universe for various deceleration parameters is schematically given in Figure 9-5. Because it would probably be confusing and ambiguous to talk about the "size" of the universe, we instead speak of the *scale of the universe.* The scale of the universe is simply any very large distance across space, such as the distance between two widely separated clusters of galaxies. Figure 9-5 shows how this very large distance changes with time, depending on the size of the deceleration parameter. If the deceleration parameter equals zero, the universe expands forever with absolutely no deceleration at all. In this case, the age of the universe must be 20 billion years. This age is deduced by simply extrapolating backward through time, assuming that the expansion rate has not slowed down over the years.

Actually, the universe must be slightly younger than 20 billion years old. The universe is not empty. The gravitational interactions between clusters of galaxies must be producing some deceleration. Consequently, the deceleration parameter must be larger than zero, and, as indicated in Figure 9-5, the Big Bang must have occurred less than 20 billion years ago. For example, if the universe is flat ($q_0 = \frac{1}{2}$), then the age of the universe should be 13 billion years. If the universe is closed, then the Big Bang occurred less than 13 billion years ago. If the universe is open, its age is between 13 and 20 billion years.

There are several ways of estimating the size of the deceleration parameter, or an equivalent quantity such as the average density

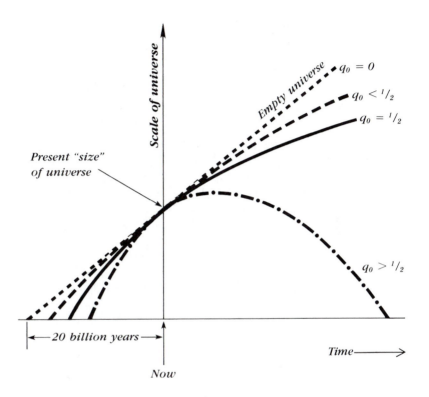

Figure 9-5 The History of the Universe
*This graph shows how the scale of the universe changes with time
for various values of the deceleration parameter (q_0).*

or curvature. Unfortunately, none of these methods is very reliable, and conflicting conclusions are often obtained.

One method involves trying to estimate the average density throughout space. By observing the motions of galaxies, we can deduce how much matter they contain. We then calculate what the average density would be if all this matter were smoothly spread across the universe. The answer is much less than the critical density. Consequently, the universe should be open and hyperbolic. The expansion should continue forever with vigor.

A second method involves trying to guess the age of the universe. This can be done by searching for the oldest stars (such as those in globular clusters) and trying to determine their ages. Alternatively, it is possible to estimate the age of the universe by measuring the relative abundances of certain ancient radioactive isotopes. From both of these approaches, we find that the age of the universe is between 10 and 18 billion years. This is disappointingly inadequate. It does not allow us to decide between an open or closed cosmology.

A third method involves measuring the amount of deuterium left over from the creation of the cosmos. Deuterium is an isotope of hydrogen. It is often called "heavy hydrogen" and constitutes an important stepping-stone in the chain of thermonuclear reactions that lead from hydrogen to helium. As I mentioned in Chapter 6, some of the primordial hydrogen was converted into helium soon after the Big Bang. During this process, some deuterium was left over; it got left behind in the chain of events that led from light hydrogen to helium when the cosmic thermonuclear reactions turned off. The amount of this primordial deuterium is very sensitive to the density and rate of expansion during the early universe. Measurements of the abundance of deuterium from satellite observations strongly suggest that the average density is much less than the critical density. Once again, this means that we live in an open, hyperbolic universe.

Yet another method involves trying to measure the deceleration of the universe directly. This is done by measuring the redshifts and distances to a large number of distant galaxies. These data are then plotted on a Hubble diagram, such as Figure 5-6. If there has been a large amount of deceleration, then the universe must have been expanding faster in the past than it is today. If there has been a small amount of deceleration, then the universe must have been expanding at a rate comparable to today's rate. These differences show up by extending the Hubble diagram to include data from the most remote galaxies, as shown in Figure 9-6.

This method should give a direct determination of the deceleration parameter. Unfortunately, like all the other methods, there is considerable uncertainty in the observations, as evidenced by the scatter of the data points on the graph in Figure 9-6.

The most courageous recent attempt to measure the deceleration parameter by this method was made by Jerome Kristian, Allan Sandage, and James Westphal at Hale Observatories. Their data, published in 1978, suggest a value of $q_0 = 1.6$. This is far above the critical value for a flat universe ($q_0 = \frac{1}{2}$) and suggests that we live in a closed, positively curved universe. Indeed, the present age of the universe should be only 12 billion years, and the expansion of the universe will stop in about 60 billion years. This would be followed by a collapse of the universe and another Big Bang roughly 130 billion years from now.

Although this is the most careful attempt to directly measure q_0 to date, Kristian, Sandage, and Westphal caution us against taking their results too seriously. For example, when you observe a galaxy that is 2 billion light years away, you are seeing how that galaxy looked 2 billion years ago. Unfortunately, we do not understand how galaxies evolve. Perhaps they were brighter in the past than they are now. Or perhaps dimmer. In either case, estimates of the distances to these remote galaxies could be in error. The conclusion that $q_0 = 1.6$ assumes that the luminosities of elliptical galaxies have not changed in at least 4 billion years.

Observations of the most distant galaxies usually focus on large elliptical galaxies in rich clusters. Quite simply, these are among the brightest and most distant galaxies we can find. But several astronomers have recently argued that the giant elliptical galaxies residing at the centers of rich clusters engage in *galactic cannibalism*. Because of their central location in the cluster and their strong gravi-

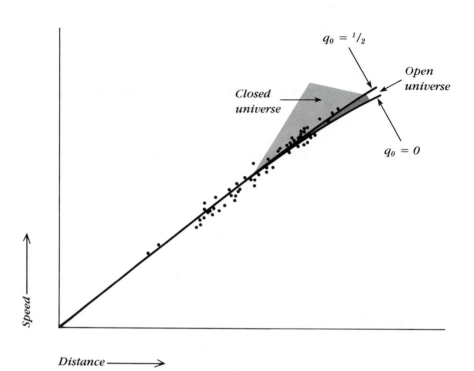

Figure 9-6   Hubble Curves and Deceleration Parameters
*It should be possible to deduce the size of the deceleration
parameter from measurements of the redshifts and distances of the
most remote galaxies we can find. If the data fall between the
curves marked $q_0 = \frac{1}{2}$ and $q_0 = 0$, then we live in an open universe
that will expand forever. If the data fall above the curve marked
$q_0 = \frac{1}{2}$, then we live in a closed universe. Unfortunately, the data
do not clearly favor either case. (Adapted from A. Sandage.)*

tational fields, these large ellipticals literally swallow up smaller galaxies that wander too near. In this way, a large elliptical galaxy becomes even larger. This perhaps explains why the giant ellipticals that dominate the centers of rich clusters always look so much bigger than their neighbors, as shown in Figure 9-7. Of course, cannibalism would change the luminosities of the large elliptical galaxies that are used in the measurements of the deceleration parameter.

The issue of the size of the deceleration parameter — and thus the ultimate future of the entire universe — obviously remains unresolved. The observations are extremely difficult because we must focus our attention on the most distant galaxies we can find in order to detect any hint of a slowing in the expansion rate of the universe. Unfortunately, at these vast distances, we are sometimes not quite sure what we are looking at. Ordinary galaxies have faded from view. Only the brightest galaxies stand out, and many of them are unusual.

When quasars were first discovered, many astronomers had high hopes that these high-redshift objects could be used as cosmic probes to explore the most remote reaches of the universe. Quasars are billions of light years away, at distances where the deceleration of the universal expansion should be easily noticeable. Unfortunately, the situation turned out to be rather complicated. Quasars come with a wide range of intrinsic luminosities. Although all quasars are bright, some are a lot brighter than others. With ordinary galaxies, it is reasonable to say that all galaxies of a particular type (for example, all Sb spirals) have roughly the same intrinsic luminosity. Thus the brightness of a galaxy is a direct measure of its distance: the bright ones are nearby, whereas the ones that look dim are farther away. No such straightforward conclusions are possible for quasars.

In 1978, several astronomers began announcing observations indicating that the strengths of certain emission lines in the spectrum of a quasar may be an excellent indicator of the quasar's true luminosity. This would enable us to distinguish the high-luminosity quasars from the low-luminosity ones. This is like being able to distinguish elliptical galaxies from spirals, or apples from oranges.

Using these important insights about the true luminosities of quasars, a team of astronomers at Lick Observatory has attempted to

Figure 9-7   Galactic Cannibalism?
Some astronomers believe that giant elliptical galaxies gravi-
tationally devour smaller galaxies in rich clusters. In this way,
these giant galaxies become even more gigantic. This photograph
shows the central regions of the Coma cluster. A wide-angle view of
this same cluster is shown in Figure 5-1. (Kitt Peak National
Observatory.)

measure the deceleration parameter. Their data favor a high value for q_0. Perhaps someday in the distant future the expansion will stop. Redshifts will turn into blueshifts as the galaxies start their headlong plunge back to a state of infinite density. In this case, the entire universe lies inside an all-encompassing black hole, and all matter and energy will be crushed out of existence as a new universe is created from the ashes of the old. And the death of our universe shall be the birth of the next.

Epilogue

I am often asked to give some sort of justification for basic research in astronomy and astrophysics. People often wonder about the wisdom of landing spacecraft on Mars or dropping probes into the Venusian atmosphere while starvation, disease, and poverty still abound in many locations here on Earth. What good is it, they ask, to study the stars? Isn't it immoral to ponder over quasars while underprivileged children are subjected to malnutrition and the skies over our cities reek with pollutants? Isn't this so-called basic research just a waste of resources, especially when the same money and talented people could be put to work on solving practical problems here at home?

There are, first of all, some economic considerations. Feats such as landing a person on the moon, or digging a trench in the Martian soil, or flying past the rings of Saturn are so dramatic and impressive that some people are misled to believe that large sums of money are involved. Obviously, the costs run into the hundreds of millions of dollars. But these amounts are insignificant when compared with the budgets for defense or welfare or many other things that the American public purchases. For example, the total cost of landing a Viking spacecraft on Mars in 1976 was $660 million. In that same year, Americans spent $570 million on pickles and $800 million on chewing gum.

It is also important to realize that we live in a science-based society. Our prosperity is entirely a result of our industry, which is based on our technology, which in turn is founded upon science. Science has become the true source of all our wealth and well-being. Without basic research, advances in science would be few and far between. Technology would stagnate, and economic growth — the only true hope for less fortunate people on our planet — would be slowed to a snail's pace.

Of course, these arguments make sense for certain areas of research. The benefits of transistors, vaccines, and high-protein grains are obvious to everyone. But what could ever come from studying galaxies and quasars?

Basic research would not be "basic" if we knew ahead of time where it would lead. For example, during the nineteenth century, Michael Faraday was asked if his discovery of electromagnetic induction would ever be of any use to humanity. He responded only by asking, "What good is a newborn baby?" At that time, it was totally unimaginable that Faraday had found the basic principle by which all electric motors and generators work.

At the turn of the twentieth century, scientists realized that they faced a perplexing problem. They were at a loss to answer the childlike question, "Why does the sun shine?" It had become clear that all their physics could not explain how the sun produces energy. But during the next two decades, many important advances occurred in nuclear physics, special relativity (with the famous equation $E = mc^2$), and quantum mechanics. By the 1920s, science had progressed to the point that Sir Arthur Eddington could argue that the sun's center is really a thermonuclear furnace in which hydrogen is fused into helium. At the center of a star, temperatures and densities are so enormous that nature can tap one of the most powerful sources of energy in the universe.

Following these revelations, Eddington was asked if humanity could ever tap this same source of energy here on Earth. Might it someday be possible to reproduce the sun's nuclear fires on our planet? Absolutely not! The absurdity of such questions was apparent to any scientist who really understood the physics of the sun's center. It was inconceivable that, in only four decades, the balance of international power would focus entirely on thermonuclear weaponry. Hydrogen bombs are devices that reproduce the sun's fire. They now dominate international politics.

Because of an unfortunate legacy involving a "caveman mentality," our knowledge of thermonuclear reactions has been put only to destructive use. Nevertheless, many scientists are now working diligently to create controlled thermonuclear fusion. If successful, we would have a virtually limitless supply of clean energy.

In the 1960s, astronomers realized that they faced a perplexing problem. They were at a loss to explain why quasars shine so brightly. Standard ideas along with the physics of the 1950s cannot explain how the luminosity of a hundred galaxies can be created in a volume no bigger than our solar system.

Almost two decades have passed since Maarten Schmidt first unraveled the spectrum of 3C273. Many important advances have been made in relativity theory. And today it seems quite clear that quasars are powered by accretion onto supermassive black holes.

Could humanity ever possibly tap the enormous energy associated with the gravitational field of a black hole? Might it someday be possible to reproduce a quasar's source of energy here on Earth? Absolutely not! The absurdity of such a suggestion is surely obvious to anyone who has the slightest understanding of quasars. Or is it?

Knowledge of the nature of physical reality bestows awesome power. To understand the intimate secrets of atoms and galaxies is to become like the gods. And we fly to the moon, light the fires of stars, and perhaps someday probe a black hole. Whether we use these abilities for the betterment of humanity or the devastation of our planet is entirely a matter of our free choice. The profound laws of nature are not evil. Only our intentions and motivations are malevolent. This is why it is so important—especially in a democratic society—to have an informed public. Only with an enlightened electorate that is aware of progress at the frontiers of science do we stand a chance of making intelligent decisions. The scientist who publishes some new discovery—however small—does so with the fervent hope that it will be used to the advantage of humanity. Whether or not these hopes are fulfilled is usually determined by politics, business, the military—and ourselves.

209

For Further Reading

For a survey of modern astronomy, you might wish to consult an elementary college text. Six excellent college texts are listed below.

Exploration of the Universe, 3rd ed., George O. Abell (Holt, Rinehart and Winston, New York, 1975).

Exploring the Cosmos, 2nd ed., Louis Berman and J. C. Evans (Little, Brown, Boston, 1977).

Astronomy: The Structure of the Universe, William J. Kaufmann (Macmillan, New York, 1977).

The Universe Unfolding, Ivan R. King (W. H. Freeman and Company, San Francisco, 1976).

Contemporary Astronomy, Jay M. Pasachoff (Saunders, Philadelphia, 1977).

Astronomy: The Evolving Universe, 2nd ed., Michael Zeilik (Harper & Row, New York, 1979).

To pursue selected topics in extragalactic astronomy and cosmology, you might want to consult the following books.

Cosmology Now, Laurie John, ed. (Taplinger, New York, 1976).

Relativity and Cosmology, 2nd ed., William J. Kaufmann (Harper & Row, New York, 1977).

Exploring the Galaxies, Simon Mitton (Scribner's, New York, 1976).

Modern Cosmology, D. W. Sciama (Cambridge University Press, Cambridge, 1971).

The Unity of the Universe, D. W. Sciama (Faber and Faber, London, 1959).

The First Three Minutes, Steven Weinberg (Basic Books, New York, 1977).

During recent years, a number of exceptional articles written for the layperson have appeared in popular journals. Some of the best are listed below, and they are grouped according to subject.

Our Galaxy
"The Center of the Galaxy," R. H. Sanders and G. T. Wrixon, *Scientific American,* vol. 230, no. 4, pp. 66–77 (April 1974).

"Steps Toward Understanding the Large Scale Structure of the Milky Way—Part I," Harold Weaver, *Mercury,* vol. 4, no. 5, pp. 18–24 (September–October 1975).

"Steps Toward Understanding the Large Scale Structure of the Milky Way—Part II," Harold Weaver, *Mercury,* vol. 4, no. 6, pp. 18–29 (November–December 1975).

Clusters of Galaxies
"Rich Clusters of Galaxies," Paul Forenstein and Wallace Tucker, *Scientific American,* vol. 239, no. 5, pp. 110–128 (November 1978).

"The Clustering of Galaxies," Edward J. Groth, James E. Peebles, Michael Seldner, and Raymond M. Soneira, *Scientific American,* vol. 237, no. 5, pp. 76– 98 (November 1977).

"Violent Tides Between Galaxies," Alar Toomre and Juri Toomre, *Scientific American,* vol. 229, no. 6, pp. 38–48 (December 1973).

Quasars and Exploding Galaxies

"In Search of X-Ray Quasars," Hale Bradt and Bruce Margon, *Sky and Telescope,* vol. 56, no. 6, pp. 499– 503 (December 1978).

"BL Lacertae Objects," Michael J. Disney and Philippe Véron, *Scientific American,* vol. 237, no. 2, pp. 32– 39 (August 1977).

"The Structure and Evolution of NGC 5128," Reginald J. Dufour and Sidney van der Bergh, *Sky and Telescope,* vol. 56, no. 5, pp. 389– 395 (November 1978).

"Exploding Galaxies," Donald Goldsmith, *Mercury,* vol. 6, no. 1, pp. 2– 5 (January– February 1977).

"Exploding Galaxies and Supermassive Black Holes," William J. Kaufmann, *Mercury,* vol. 7, no. 5, pp. 97– 105 (September– October 1978).

"Quasars in Ultraviolet," Dennis Overbye, *Sky and Telescope,* vol. 55, no. 1, pp. 31– 33 (January 1978).

"Quasi-Stellar Objects," Harding E. Smith, *Mercury,* vol. 7, no. 2, pp. 27– 33 (March– April 1978).

"Giant Radio Galaxies," Richard G. Strom, George K. Miley, and Jan Oort, *Scientific American,* vol. 233, no. 2, pp. 26– 35 (August 1975).

Cosmology and the Universe

"Cosmology—The Origin and Evolution of the Universe," George O. Abell, *Mercury,* vol. 7, no. 3, pp. 45– 61 (May–June 1978).

"The Curvature of Space in a Finite Universe," J. J. Callahan, *Scientific American,* vol. 235, no. 2, pp. 90– 100 (August 1976).

"Will the Universe Expand Forever?" J. Richard Gott III, James E. Gunn, David N. Schramm, and Beatrice M. Tinsley, *Scientific American,* vol. 234, no. 3, pp. 62–79 (March 1976).

"Cosmology Today," Louis C. Green, *Sky and Telescope,* vol. 54, no. 3, pp. 180–184 (September 1977).

"The Future History of the Universe," John K. Lawrence, *Mercury,* vol. 7, no. 6, pp. 132–138 (November–December 1978).

"The Arrow of Time," David Layzer, *Scientific American,* vol. 233, no. 6, pp. 56–69 (December 1975).

"The Cosmic Background Radiation and the New Aether Drift," Richard A. Muller, *Scientific American,* vol. 238, no. 5, pp. 64–75 (May 1978).

"Deuterium in the Universe," Jay M. Pasachoff and William A. Fowler, *Scientific American,* vol. 230, no. 5, pp. 108–118 (May 1974).

"The Age of the Elements," David N. Schramm, *Scientific American,* vol. 230, no. 1, pp. 69–77 (January 1974).

"The Cosmic Background Radiation," Adrian Webster, *Scientific American,* vol. 231, no. 2, pp. 26–33 (August 1974).

Appendix:

The Messier Catalogue

Messier number	NGC number	Position (1950) R.A. h	R.A. m	Decl. °	Decl. '	Constellation	Distance (light years)	Comments
M1	1952	5	31.5	+21	59	Taurus	6,000	supernova remnant
M2	7089	21	30.9	− 1	03	Aquarius	50,000	globular cluster
M3	5272	13	39.9	+28	38	Canes Venatici	30,000	globular cluster
M4	6121	16	20.6	−26	24	Scorpius	10,000	globular cluster
M5	5904	15	16.0	+ 2	16	Serpens	30,000	globular cluster
M6	6405	17	36.8	−32	11	Scorpius	2,000	open cluster
M7	6475	17	50.7	−34	48	Scorpius	1,000	open cluster
M8	6523	18	01.6	−24	20	Sagittarius	6,500	nebula
M9	6333	17	16.2	−18	28	Ophiuchus	25,000	globular cluster
M10	6254	16	54.5	− 4	02	Ophiuchus	16,000	globular cluster
M11	6705	18	48.4	− 6	20	Scutum	6,000	open cluster
M12	6218	16	44.6	− 1	52	Ophiuchus	16,000	globular cluster
M13	6205	16	39.9	+36	33	Hercules	25,000	globular cluster
M14	6402	17	35.0	− 3	13	Ophiuchus	23,000	globular cluster
M15	7078	21	27.6	+11	57	Pegasus	40,000	globular cluster
M16	6611	18	16.0	−13	48	Serpens	7,000	nebula and cluster
M17	6618	18	18.0	−16	12	Sagittarius	5,000	nebula and cluster
M18	6613	18	17.0	−17	09	Sagittarius	6,000	open cluster
M19	6273	16	59.5	−26	11	Ophiuchus	20,000	globular cluster
M20	6514	17	58.9	−23	02	Sagittarius	2,200	nebula

M21	6531	18	01.8	−22	30	Sagittarius	3,000	open cluster
M22	6656	18	33.3	−23	58	Sagittarius	10,000	globular cluster
M23	6494	17	54.0	−19	01	sagittarius	4,500	open cluster
M24	6603	18	15.5	−18	27	Sagittarius	10,000	star cloud
M25 IC	4725	18	28.8	−19	17	Sagittarius	2,000	open cloud
M26	6694	18	42.5	− 9	27	Scutum	5,000	open cluster
M27	6853	19	57.4	+22	35	Vulpecula	1,250	planetary nebula
M28	6626	18	21.5	−24	54	Sagittarius	15,000	globular cluster
M29	6913	20	22.2	+38	21	Cygnus	7,200	open cluster
M30	7099	21	37.5	−23	25	Capricornus	40,000	globular cluster
M31	224	00	40.0	+41	00	Andromeda	2,200,000	spiral galaxy
M32	221	00	40.0	+40	36	Andromeda	2,200,000	elliptical galaxy
M33	598	1	31.1	+30	24	Triangulum	2,300,000	spiral galaxy
M34	1039	2	38.8	+42	34	Perseus	1,400	open cluster
M35	2168	6	05.7	+24	20	Gemini	2,800	open cluster
M36	1960	5	32.0	+34	07	Auriga	4,100	open cluster
M37	2099	5	49.0	+32	33	Auriga	4,600	open cluster
M38	1912	5	25.3	+35	48	Auriga	4,200	open cluster
M39	7092	21	30.4	+48	13	Cygnus	900	open cluster
M40	none	12	20.0	+58	22	Ursa Major		double star
M41	2287	6	44.9	−20	42	Canis Major	2,400	open cluster
M42	1976	5	32.9	− 5	25	Orion	1,000	nebula
M43	1982	5	33.1	− 5	18	Orion	1,000	nebula
M44	2632	8	37.5	+19	52	Cancer	500	open cluster
M45	none	3	43.9	+23	58	Taurus	400	open cluster

(continued)

Messier number	NGC number	Position (1950) R.A. h	m	Decl. °	′	Constellation	Distance (light years)	Comments
M46	2437	7	39.6	−14	42	Puppis	5,400	open cluster
M47	2422	7	34.3	−14	22	Puppis	1,600	open cluster
M48	2548	8	11.2	−5	38	Hydra	1,500	open cluster
M49	4472	12	27.3	+8	16	Virga	70,000,000	elliptical galaxy
M50	2323	7	00.5	−8	16	Monoceros	3,000	open cluster
M51	5194	13	27.8	+47	27	Canes Venatici	15,000,000	spiral galaxy
M52	7654	23	22.0	+61	20	Cassiopeia	7,000	open cluster
M53	5024	13	10.5	+18	26	Coma Berenices	60,000	globular cluster
M54	6715	18	52.0	−30	32	Sagittarius	50,000	globular cluster
M55	6809	19	36.9	−31	03	Sagittarius	20,000	globular cluster
M56	6779	19	14.6	+30	05	Lyra	40,000	globular cluster
M57	6720	18	51.7	+32	58	Lyra	4,100	planetary nebula
M58	4579	12	35.1	+12	05	Virgo	70,000,000	barred spiral galaxy
M59	4621	12	39.5	+11	55	Virgo	70,000,000	elliptical galaxy
M60	4649	12	41.1	+11	49	Virgo	70,000,000	elliptical galaxy
M61	4303	12	19.4	+4	45	Virgo	70,000,000	barred spiral galaxy
M62	6266	16	58.1	−30	03	Ophiuchus	26,000	globular cluster
M63	5055	13	13.5	+42	17	Canes Venatici	14,500,000	spiral galaxy
M64	4826	12	54.3	+21	57	Coma Berenices	12,000,000	spiral galaxy
M65	3623	11	16.3	+13	23	Leo	35,000,000	spiral galaxy

	NGC	h	m	°	'			
M66	3627	11	17.6	+13	17	Leo	35,000,000	spiral galaxy
M67	2682	8	47.8	+12	00	Cancer	2,250	open cluster
M68	4590	12	36.8	−26	29	Hydra	40,000	globular cluster
M69	6637	18	28.1	−32	33	Sagittarius	25,000	globular cluster
M70	6681	18	40.0	−32	21	Sagittarius	65,000	globular cluster
M71	6838	19	51.5	+18	39	Sagitta	8,500	globular cluster
M72	6981	20	50.7	−12	44	Aquarius	60,000	globular cluster
M73	6994	20	56.4	−12	50	Aquarius		cluster of four stars
M74	628	1	34.0	+15	32	Pisces	20,000,000	spiral galaxy
M75	6864	20	03.2	−22	04	Sagittarius	100,000	globular cluster
M76	650	1	38.8	+51	19	Perseus	3,400	planetary nebula
M77	1068	2	40.1	− 0	14	Cetus	30,000,000	Seyfert galaxy
M78	2068	5	44.2	+ 0	02	Orion	1,600	nebula
M79	1904	5	22.2	−24	34	Lepus	54,000	globular cluster
M80	6093	16	14.1	−22	52	Scorpius	36,000	globular cluster
M81	3031	9	51.5	+69	18	Ursa Major	7,000,000	spiral galaxy
M82	3034	9	51.9	+69	56	Ursa Major	7,000,000	irregular galaxy
M83	5236	13	34.3	−29	37	Hydra	8,000,000	spiral galaxy
M84	4374	12	22.6	+13	10	Virgo	70,000,000	elliptical galaxy
M85	4382	12	22.8	+18	28	Coma Berenices	70,000,000	elliptical galaxy
M86	4406	12	23.7	+13	13	Virgo	70,000,000	giant elliptical galaxy
M87	4486	12	28.3	+12	40	Virgo	70,000,000	giant elliptical galaxy
M88	4501	12	29.5	+14	42	Coma Berenices	40,000,000	spiral galaxy
M89	4552	12	33.1	+12	50	Virgo	70,000,000	elliptical galaxy
M90	4569	12	34.3	+13	26	Virgo	70,000,000	spiral galaxy

(continued)

Messier number	NGC number	Position (1950) R.A. h	m	Decl. °	'	Constellation	Distance (light years)	Comments
M91	4548	12	32.9	+14	46	Coma Berenices	40,000,000	barred spiral galaxy
M92	6341	17	15.6	+43	12	Hercules	28,000	globular cluster
M93	2447	7	42.4	−23	45	Puppis	36,000	open cluster
M94	4756	12	48.6	+41	23	Canes Venatici	14,500,000	spiral galaxy
M95	3351	10	41.3	+11	58	Leo	25,000,000	barred spiral galaxy
M96	3368	10	44.2	+12	05	Leo	25,000,000	spiral galaxy
M97	3587	11	12.0	+55	18	Ursa Major	2,600	planetary nebula
M98	4192	12	11.3	+15	11	Coma Berenices	70,000,000	spiral galaxy
M99	4254	12	16.3	+14	42	Coma Berenices	70,000,000	spiral galaxy
M100	4321	12	20.4	+16	06	Coma Berenices	70,000,000	spiral galaxy
M101	5457	14	01.4	+54	35	Ursa Major	15,000,000	spiral galaxy
M102 (= M101)								
M103	581	1	29.9	+60	27	Cassiopeia	8,000	open cluster
M104	4594	12	37.3	−11	21	Virgo	50,000,000	spiral galaxy
M105	3379	10	45.2	+12	51	Leo	25,000,000	elliptical galaxy
M106	4258	12	16.5	+47	35	Canes Venatici	25,000,000	spiral galaxy
M107	6171	16	29.7	−12	57	Ophiuchus	10,000	globular cluster
M108	3556	11	08.7	+55	57	Ursa Major	25,000,000	spiral galaxy
M109	3992	11	55.0	+53	39	Ursa Major	25,000,000	barred spiral galaxy
M110	205	00	37.6	+41	25	Andromeda	2,200,000	small elliptical galaxy

Index